Sustainable Development Goals Series

Good Health and Well-being

The **Sustainable Development Goals Series** is Springer Nature's inaugural cross-imprint book series that addresses and supports the United Nations' seventeen Sustainable Development Goals. The series fosters comprehensive research focused on these global targets and endeavours to address some of society's greatest grand challenges. The SDGs are inherently multidisciplinary, and they bring people working across different fields together and working towards a common goal. In this spirit, the Sustainable Development Goals series is the first at Springer Nature to publish books under both the Springer and Palgrave Macmillan imprints, bringing the strengths of our imprints together.

The Sustainable Development Goals Series is organized into eighteen subseries: one subseries based around each of the seventeen respective Sustainable Development Goals, and an eighteenth subseries, "Connecting the Goals," which serves as a home for volumes addressing multiple goals or studying the SDGs as a whole. Each subseries is guided by an expert Subseries Advisor with years or decades of experience studying and addressing core components of their respective Goal.

The SDG Series has a remit as broad as the SDGs themselves, and contributions are welcome from scientists, academics, policymakers, and researchers working in fields related to any of the seventeen goals. If you are interested in contributing a monograph or curated volume to the series, please contact the Publishers: Zachary Romano [Springer; zachary.romano@springer.com] and Rachael Ballard [Palgrave Macmillan; rachael.ballard@palgrave.com].

Eucharia Oluchi Nwaichi
Editors

Modernity in Health and Disease Diagnosis: The Account from STEM Women

Springer

Editor
Eucharia Oluchi Nwaichi
Department of Biochemistry, Faculty of Science, University of Port Harcourt
Organization for Women in Science for the Developing World (OWSD)
University of Port Harcourt Branch; and Exchange & Linkage Programmes Unit
University of Port Harcourt
Port Harcourt, Nigeria

ISSN 2523-3084 ISSN 2523-3092 (electronic)

ISBN 978-3-031-34962-1 ISBN 978-3-031-34963-8 (eBook)
https://doi.org/10.1007/978-3-031-34963-8

This Springer imprint is published by the registered company Springer Nature Switzerland AG
The registered company address is: Gewerbestrasse 11, 6330 Cham, Switzerland

Dedicated to the trailblazing women in STEM who have paved the way for future generations, and to all those who continue to push the boundaries of modern medicine with their unwavering curiosity, innovative thinking, and tireless pursuit of better health and disease diagnosis for all. Your contributions to this field are immeasurable, and this book is a testament to the profound impact you have had on the world. Thank you for inspiring us to always strive for progress and for reminding us that with hard work, determination, and a passion for knowledge, anything is possible.

Foreword

I feel privileged to write the foreword of this book *Modernity in Health and Disease Diagnosis: The Account from STEM Women* compiled by Prof. Eucharia Nwaichi and her team. I met Prof. Eucharia Nwaichi through the AuthorAID community where we both belong and I have always admired her passion for research excellence. The University of Port Harcourt Branch of the Organization for Women in Science for the Developing World (OWSD), where she leads in the capacity of Chair, is known to be one of the very active and productive OWSD Branches in Nigeria. It is therefore not unexpected that a book of this quality is being written by the members of the Branch, who are having major impact on science, technology, engineering, and math (STEM).

This book is relevant to the Sustainable Development Goals 3, 5, and 10. As a researcher, environmental toxicologist, and SDG advocate, I understand how essential this book will be to fellow researchers and practitioners working in the SDG space. I am excited that all the main authors are women scientists, in an era where advancement of women in STEM fields is being encouraged, they are making meaningful contribution in the field of science through this book. These female authors have demonstrated excellence in research, cutting across different areas of the sciences ranging from technology, women's health, environmental sciences, and food security amongst others. I believe this book will be an inspiration to other women scientists and female students, facing challenges on how to balance family, work, and be able to still make notable impact and advance in their chosen careers.

Funmilayo Doherty
AuthorAID National Research Hub
Lagos, Nigeria

Society for Environmental Toxicology
and Pollution Mitigation (SETPOM)
Lagos, Nigeria

Yaba College of Technology
Lagos, Nigeria

Acknowledgments

I am deeply grateful for the support received from my amazing fan and husband, Chucks, and our lovely twins Francis Munachimso and Claire Somtochukwu. From serving warm reminders to setting up virtual meetings where most presentations were made before conception as chapters, he was as important to this book getting done as I was. Muna and Somto (in their less than 2 years of age) were patient and kind with me in my late arrivals from the office, working tirelessly on the book.

Thanks to everyone on the Books Series Committee of Organization for Women in Science University of Port Harcourt (OWSD UniPort) who helped in various ways, chiefly the Chair—Prof. A. I. Hart, Dr. B. Odogwu (Secretary), Dr. E. Fomsi, Dr. C. C. Ikewuchi, Dr. M. T. Nwakuya, and Mrs. Martha Musa.

The Vice Chancellor of the University of Port Harcourt, Prof. Owunari Abraham Georgewill, Prof. Nimi D. Briggs (late now), Prof. Onyewuchi Akaranta, and Prof. Joseph Ajienka are warmly appreciated for their encouragement and support throughout the project. Prof. Florence Nduka, Prof. Helen Imafido, Prof. E. O. Nwaichi, Dr. Ifeoma Chidebe, Dr. Ronald Kakeeto, Dr. Sally Chikuta, Dr. Eric Agoyi, Dr. Nwakuya, and Dr. C. C. Ikewuchi are highly appreciated for painstakingly reviewing assigned chapters of this book.

The authors of this book would like to express their sincere gratitude to all the inspiring and dedicated STEM women who shared their experiences and insights with us. Their contribution was invaluable and made this project possible.

The chapter contributors are awesome. Their passion and commitment to excellence made this a truly collaborative and rewarding experience.

Finally, we would like to acknowledge the support of Mary James for prompt responses to my endless queries, our publisher and editors, who believed in this project from the start and helped bring it to fruition.

Contents

1 **Introduction** 1
Elsie I. Hamadina

2 **Paradigm Shift in Health-Related Academic Research with Cloud Computing** 7
Ugochi Adaku Okengwu

3 **Significance of Nutritional Etiquette to Women's Health** 19
Ododobari Jike-Wai

4 **Women at Greater Risk of Alzheimer's: Way Forward** 29
Chinagorom Petrolina Ibeachu

5 **Leveraging Artificial Intelligence Technology for Effective Early Diagnosis: Heart Issues** 41
Laeticia Nneka Onyejegbu

6 **Antimalarial Drug Resistance and Vulnerable Groups** 49
Ifeyinwa N. Chijioke-Nwauche and Amaka M. Awanye

7 **Enhancing Healthcare Through Automation and Robotics** 59
Nkolika Ogechukwu Nwazor and Stella Ifeoma Orakwue

8 **Naturally Derived Surfactants for Healthy Food Formulation** 69
Millicent Uzoamaka Ibezim-Ezeani

9 **Rethinking Agenda 2063: Leveraging STEM Women's Empowerment for Food Security in a Post-COVID-19 Pandemic Era** 83
A. Henri-Ukoha and I. I. Ukoha

10 **Antimicrobial Resistance: A Collective Responsibility** 89
Kome Otokunefor, Obakpororo Ejiro Agbagwa, Esther Oshifade, Chidozie Ngozi Endarlyn Ibezim, and Esther Azi

11 **Accessing Library Information Services in Public Health Emergencies** 99
Helen Uzoezi Emasealu

12 Human Gut Microbiome: Its Role in Health and Development 107
Obakpororo Ejiro Agbagwa

13 Chemical Safety and Chemical Security: A Guide to Preventing Health Hazards 117
Kaine Okorosaye-Orubite

14 Chemical Leaching into Food and the Environment Poses Health Hazards 129
Gloria Ukalina Obuzor and Uchechi Bliss Onyedikachi

List of Contributors

Obakpororo Ejiro Agbagwa Department of Microbiology, Faculty of Science, University of Port Harcourt, Port Harcourt, Rivers State, Nigeria

Amaka M. Awanye Department of Pharmaceutical Microbiology and Biotechnology, Faculty of Pharmaceutical Sciences, University of Port Harcourt, Port Harcourt, Rivers State, Nigeria

Esther Azi Department of Community Medicine, University of Port Harcourt Teaching Hospital, Port Harcourt, Nigeria

Ifeyinwa N. Chijioke-Nwauche Department of Clinical Pharmacy and Management, Faculty of Pharmaceutical Sciences, University of Port Harcourt, Port Harcourt, Rivers State, Nigeria

Helen Uzoezi Emasealu Department of Library and Information Science, University of Harcourt, Choba, Rivers State, Nigeria

Elsie I. Hamadina Department of Crop and Soil Science, University of Port Harcourt, Port Harcourt, Nigeria

A. Henri-Ukoha Department of Agricultural Economics and Agribusiness Management, University of Port Harcourt, Choba, Rivers State, Nigeria

Chinagorom Petrolina Ibeachu Department of Anatomy, Faculty of Basic Medical Sciences, University of Port-Harcourt, Port Harcourt, Nigeria

Chidozie Ngozi Endarlyn Ibezim Department of Pharmaceutical Microbiology and Biotechnology, Faculty of Pharmaceutical Sciences, University of Port Harcourt, Port Harcourt, Nigeria

Millicent Uzoamaka Ibezim-Ezeani Department of Pure and Industrial Chemistry, University of Port Harcourt, Choba, Rivers State, Nigeria

Ododobari Jike-Wai Department of Food, Nutrition and Home Science, Faculty of Agriculture, University of Port Harcourt, Port Harcourt, Rivers State, Nigeria

Nkolika Ogechukwu Nwazor Electrical/Electronic Engineering Department, University of Port Harcourt, Port Harcourt, Nigeria

Gloria Ukalina Obuzor Department of Pure and Applied Chemistry, University of Port Harcourt, Port Harcourt, Nigeria

Ugochi Adaku Okengwu Department of Computer Science, University of Port Harcourt, Port Harcourt, Rivers State, Nigeria

Uchechi Bliss Onyedikachi Department of Biochemistry, College of Natural Sciences, Michael Okpara University of Agriculture, Umudike, Abia State, Nigeria

Laeticia Nneka Onyejegbu Computer Science Department, Faculty of Computing, University of Port Harcourt, Port Harcourt, Rivers State, Nigeria

Stella Ifeoma Orakwue Electrical/Electronic Engineering Department, University of Port Harcourt, Port Harcourt, Nigeria

Kaine Okorosaye Orubite Department of Pure and Applied Chemistry, University of Port Harcourt, Port Harcourt, Nigeria

Esther Oshifade Department of Animal Science, Faculty of Agriculture, University of Port Harcourt, Port Harcourt, Nigeria

Kome Otokunefor Department of Microbiology, Faculty of Science, University of Port Harcourt, Port Harcourt, Nigeria

I. I. Ukoha Department of Agricultural Economics, Federal University of Technology, Owerri, Imo State, Nigeria

1 Introduction

Elsie I. Hamadina

From the agrarian era through to the industrial era and now the electronic/information era, the subject of human health and disease diagnosis has remained a significant part of the central concerns of man in his determination for worthwhile survival on Earth. What has changed over these periods, however, are the definition of human health; the kind and extent to which technology is applied to problem-solving; the speed/ease of acquiring, transmitting, and utilizing data for the prediction of health status or disease diagnosis; and the application of multidisciplinary approaches to health care such as the simultaneous use of knowledge from agriculture, medicine, nutrition, economics, etc., to mention a few. In ancient times, health was seen from a religious perspective as one in which good health comes from God or some deity (Badash et al., 2017). Later definitions saw the inclusion of social perspectives such as temperance and personality (Badash et al., 2017), then, to the more all-inclusive definition of good health by the World Health Organization (WHO), as "a state of complete physical, mental and social well-being, not merely the absence of disease" (WHO, 1946). Another currently accepted view on health is that suggested by Dubos (1959). His definition of health, as "the condition best suited for each person to reach his or her current personal and social goals," recognizes that a person's health status can change over time and space/environment. This makes early identification/diagnosis of a shift from complete wellness to any level of unwellness along the continuum from wellness to unwellness a key to sustaining man at the height of his potential. Modernity in health and disease diagnosis, therefore, has given man the impetus to aspire to attain a "health for all" status (i.e., physical, mental, and social wellness for all to attain their maximum potential at any one time).

The actualization of health for all across the globe is driven by the United Nations under its Sustainable Development Goals (SDGs) program and the African Union under its Aspiration 1 of Agenda 2063 program and implemented by all governments and actors signed on each of the programs. The 17 SDGs were set up in 2015 to serve as a "blueprint to achieve a better and more sustainable future for all" by the year 2030.

In Nigeria, the women in STEM in the University of Port Harcourt, under the aegis of Organization for Women in Science for the Developing World (OWSD), University of Port Harcourt Chapter, are contributing to the global pursuit of sustainable health for all through the delivery of monthly seminars around the theme "Modernity in health and disease diagnosis: The account of women in STEM" and the publication of the papers for dissemination to a range of audience across the globe. The title Modernity in

E. I. Hamadina (✉)
Department of Crop and Soil Science, University of Port Harcourt, Port Harcourt, Nigeria
e-mail: Elsie.hamadina@uniport.edu.ng

E. O. Nwaichi (ed.), *Modernity in Health and Disease Diagnosis: The Account from STEM Women*, Sustainable Development Goals Series, https://doi.org/10.1007/978-3-031-34963-8_1

Health and Disease Diagnosis: The Account from STEM women is a compilation of chapters that bring to the fore interesting insights on the current state of knowledge and best practices in the topics addressed. The topics can be broadly grouped into two: those that cover the application of current computer technology in early disease diagnosis, treatment and research (Chaps. 2, 5 and 7) and those that present the current knowledge on human health from the perspective of agriculture, nutrition, economics, chemistry and communication (Chaps. 3, 4, 6, 8–14). They largely provide information that border on SDG 3- "to ensure healthy lives and promote well-being for all at all ages," yet some address other SDGs such as SDG 2- "End hunger, achieve food security and improved nutrition and promote sustainable agriculture," SDG 5- "Achieve gender equality and empower all women and girls," and SDG 9- "Build resilient infrastructure, promote inclusive and sustainable industrialization and foster innovation." Chapter 2 is an interesting article on Cloud Computing." Here, readers will be enraptured by the concept of cloud computing and its promising potential as they learn of the variety of opportunities for manipulation, configuration, and access to data through cloud computing–enabled resources. They will also be enlightened on basic cloud computing service models and the best ways to utilize them in various service delivery sectors as well as in academic research. Researchers and practitioners working on all SDG targets and indicator workspaces will benefit from this chapter by way of enhanced knowledge and work efficiency but more specifically to workers in the 9.4, 9.5, 9a, and 9b target workspaces. In Chap. 3 , "Significance of Nutritional Etiquette to Women's Health," the author discusses the importance of nutritional health which is directly related to how the body utilizes food nutrients for development and optimal performance, and how responsible eating choices amount to normal body dimensions (anthropometry). Readers will learn the benefits of engaging in nutritional etiquette for sound health. They will also appreciate the need to consciously build an optimum nutritional status to guard against malnutrition. The chapter further furnishes readers with nutritional behaviors that are significant to women's health and their consequent effect on fertility/reproductive health and outcomes on offspring. The chapter addresses SDG 2 and will be useful to workers in its various target and indicator workspaces. Chapter 4 provides current information on the brain disease called Alzheimer's, a progressive and degenerative brain disease that results from the stiffening of healthy brain cells to death, which in turn leads to increasing loss of memory. Readers will find detailed information on how and why this happens, and why women are at greater risk of Alzheimer's. This chapter would spur women in STEM to read and practice the lifestyle changes discussed in the text as they become apprised of the factors that predispose especially women to this chronic disease. Statistical evidence on women and Alzheimer's contained in the chapter will prove a startling need for intentional individual and collective effort toward exploring avenues for prevention or deceleration of the progression of the disease, thus, enabling women (and men) to remain helpful to society. This chapter will be useful to researchers and practitioners working on SDG 3.4.1 (C030401) target and indicator workspace as well as those in target 5.b "Enhance the use of enabling technology, in particular information and communications technology, to promote the empowerment of women." In Chap. 5, readers will learn about artificial intelligence (AI), its latest application in the early detection of heart diseases, and how such early detection saves lives and ensures a better quality of life. Readers will learn about the powerful neural network (NN)/deep learning AI model that is inspired by how the biological neuron works and how it is used for the early detection of heart failure. They will also learn about machine learning, an AI model that is made up of algorithms that can improve themselves over time using big data, and how this makes them highly suitable for solving new problems in an ever-changing world. The chapter also explains how the application of a new AI machine learning algorithm has made it possible to detect, early enough, two life-threatening heart conditions (hypertrophic cardiomyopathy and cardiac amyloidosis) that would otherwise be impossible to detect without AI. It will be evident how these, indeed, set the pace for a revolutionary diagnosis, treatment, risk prediction, and clinical care in cardiovascular medicine. This chapter will be useful to researchers and practitioners working on SDG 3.4.1 (C030401) target and indicator workspace. Chap. 6 discusses antimalarial drug resistance, and highlights women as the most vulnerable group. It is an engaging read as readers learn about various vulnerable groups to malaria, different malaria-causing plasmodium parasite species, their significant levels, etc. Therein, factors that impose vulnerability on women are discussed and backed with statistical evidence that span across a few continents. A major highlight in the chapter is the WHO's rec ommendation for the establishment of a malaria treatment that ensures the annihilation of the parasite and the factors that could antagonize the effort, resulting in treatment failure. The precursors of anti-malarial drug resistance are also addressed, and gender connections are made especially as it relates to women as they are advised on the steps to follow to deal with and combat this resistance. This chapter addresses SDG 3 and will be useful to researchers and practitioners working on SDG 3.3, and 3d target and indicator workspaces. Chap. 7, is an exposé on robotics and automation. It also explained the difference between robotics and automation and showed how these innovations have transformed on healthcare services. Readers will learn that automation has found applications in areas such as patients' and doctors' appointment scheduling, financial management, healthcare staff protection and safety, medical record-keeping, clinical diagnosis, surgical procedures, and decision-making by analyzing patient data and recommending treatments. Readers will also learn that robotics, on the other hand, have been applied in various healthcare settings. Some examples include the impact of medical robots in enhancing surgical procedures and precision tasks, and in minimizing invasiveness. Also, the roles of industrial and domestic robots are discussed, and worthy of note is the role of Unmanned Aerial Vehicles and boats in rescue missions and surveillance. This chapter also highlights the role of robotics in patient care such as in the monitoring of patients and in the development of prostheses for amputees. Additionally, the role of robotics in diagnosing and treating diseases, particularly in situations where contact between doctors and patients needs to be minimized, such as during infectious disease outbreaks are discussed. Researchers and practitioners working on SDG 3.3, 3.4, 3.6, 3.8, 3b and 3d target and indicator workspaces will find this chapter useful for monitoring and evaluation. Chap. 8, discusses the significance of naturally-derived surfactants in healthy food formulations and their role in achieving improved nutrition, food preservation and food security in a rapidly growing global population. Readers will learn that formulated foods, which are foods prepared with ingredients that enhance the nutritional value and provide additional nutraceutical benefits, are not only composed of carbohydrates, proteins, fats, or lipids, etc., but also contain naturally-derived surfactants (NDSs) that function as stabilizer through processes like emulsification and dispersion. The utilization of NDSs in various food applications, such as confectionery, emulsions, fat spreads, and bakery, is highlighted. This chapter addresses SDG 2, particularly targets 2.3 and 2a. Chap. 9, "Re-thinking Agenda 2063: Leveraging STEM Women Empowerment for Food Security in a PostSDGs and target-Covid-19 Pandemic Era," enunciates the adverse effects and socioeconomic impacts of the COVID-19 pandemic with a special focus on its impact on women. Furthermore, the distinct place of women in curtailing the magnitude of hungry people, globally, through various agricul tural enterprises that create a formidable post-COVID food system was articulated. The chapter proposes the engagement of STEM women in a strategic manner to achieve food security and alleviate the projected estimation of people who would go to bed hungry, especially following the COVID-19 pandemic, through two broad paths: (1) the direct empowerment of STEM women and (2) the transfer of knowledge to relevant beneficiaries. This chapter addresses SDG 5, particularly targets 5.5 and 5c, "to ensure women's full and effective participation and equal opportunities for leadership at all levels of decision-making in political, economic and public life" and "adopt and strengthen sound policies and enforceable legislation for the promotion of gender equality and the empowerment of all women and girls at all levels," respectively. In Chap. 10, readers will come across a discussion on the threat to life imposed by the global plague of antimicrobial resistance, the need to urgently tackle its spread, and how to tackle its spread simultaneously at the human, animal, and environmental levels. Readers will note the dangers humans are exposed to on account of the high (87%) resistance of widely available *E. coli* and ESKAPE pathogens to drugs. This chapter addresses SDG 3 and will be useful to researchers and practitioners working on SDG 3.3.3, 3.d.1, and 3.d.2 target and indicator workspaces. In Chap. 11, curious librarians and interested others will acquire information on how to systematize library information to meet public health emergencies in this fast-changing, developing world. They would become abreast with the impending dangers that could plague "Information-Vulnerable Members of Society" as they become aware of the place of knowledge and authentic information, especially in public health emergencies. The challenge of information handling in the modern world and possible mitigative measures against those challenges as well as the place of appropriate professional skill set including, broadly, sourcing, verification, identification, and dissemination of information in information handling are found therein. This chapter addresses SDG 9 and will be useful to researchers and practitioners working on SDG 9.5, 9.a, 9.b, and 9.c target workspaces and their indicators. In Chap. 12, information on the diseases associated with imbalances in the human gut microbiome is presented. Readers will learn about the microbes that reside within the human gastrointestinal tract (GIT) – collectively referred to as gut microbiome (GM). Readers will appreciate the human gut microbiomes differently after reading this chapter. They will become conversant with the development of the gut microbiome; gut microbiome roles and constituents; proportions of microbes resident in the human gut among others (as most of these microbes are beneficial while others are capable of causing diseases, malfunctioning, and altogether affecting physiological functions); their relationship with the development of diseases of the human body; and the recommended actions to keep the system balanced for healthy living, particularly for women. It is hoped that researchers and practitioners working on SDG 3.4.1 (C030401) target and indicator workspace will find this chapter useful in the planning, implementation, and monitoring of projects focused on enhancing the health status of women. In Chap. 13, the topic "Chemical Safety and Chemical Security; a Guide to Preventing Health Hazards" is discussed. The chapter posits that chemical safety and chemical security are two essential practices that minimize the adverse effects of chemicals on the environment. According to the article, Chemical safety refers to practices that protect humans and the environment from exposure to hazardous chemicals, while chemical security aims at preventing the deliberate misuse or theft of chemicals for criminal activities. Chemistry enthusiasts and professionals will enjoy this read particularly when in search for best practices to adopt while at work with chemicals and explosives or during the production of hand sanitizers as was the case during the COVID-19 pandemic era. This chapter addresses SDG 2, particularly targets 2.3 and 2c. Finally, Chap. 14, entitled "Chemical Leaching into Food and the Environment Poses Health Hazards," exposes the dangers of the misuse of plastics to humans and the environment, particularly the danger of cooking with plastics on fetus, children, and even adults. Readers will learn about toxic metalloids and important chemicals that, if ingested, could result in severe endocrinal anomalies in humans. These toxic chemicals are also persistent in the environment and are purported to drive many challenges which are mostly long-term. Readers will also learn the role of "additives" in the production of plastics and their relationship to plastic pollution. The chapter will educate readers on the classes or types of plastics and how knowledge of these class types can help consumers ward off various health and environmental challenges that could arise from the inappropriate use of these materials. It calls the attention of all to the need to use modern Plastic Codes to determine their ideal usage. This chapter addresses SDG 3 and will be useful to researchers and practitioners working on SDG 3.9.2, 3.9.3, 3.d.1, and 3.d.2 target and indicator workspaces.

References

Badash I, Kleinman NP, Barr S, Jang J, Rahman S, Wu BW. Redefining health: the evolution of health ideas from antiquity to the era of value-based care. Cureus. 2017;9(2):e1018. https://doi.org/10.7759/cureus.1018.

Dubos R. *Mirage of health: utopias, Progress, and biological change*. New York: Harper and Brothers; 1959. [Google Scholar]

WHO, 1946. https://www.who.int/governance/eb/who_constitution_en.pdf/

Paradigm Shift in Health-Related Academic Research with Cloud Computing

2

Ugochi Adaku Okengwu

2.1 Introduction

In both teaching and health-related research, information technology is critical. Cloud computing is a breakthrough in the field of information technology that allows us to use software as a utility over the internet. Cloud computing is the most recent effort in providing computing resources as a service, which has resulted in the simplicity of many academic research methodologies through the use of programming packages, resulting in the use of many applications on our PCs. The concept of cloud computing is an extension of distributed computing, which is the process of running applications across multiple computers connected by a network (Pinal 2012). It also provides information technology property and capabilities (programs, storage, communication, collaboration, and infrastructure) through CSP services (cloud service provider). Cloud computing is quickly becoming one of the most well-known and promising innovations. It provides a set of opportunities for businesses to improve their operations and make better use of technology (Colin and Felicia 2016). In traditional computing, we install software program packages on a personal computer (laptop) to replace the system as needed, but in cloud computing, the software program applications are stored on servers and accessed via the internet rather than being operated from a personal computer. Cloud computing is a relatively new and rapidly expanding trend in healthcare. New ways of producing, distributing, and utilizing services are enabled by ubiquitous, on-demand access to essentially infinite resources combined with a pay-per-use paradigm. Cloud computing is frequently employed in an OMICS context, such as for genomics, proteomics, and molecular medicine computing. Most healthcare research employs cloud computing technologies to provide extensive network connectivity, allowing for data sharing, data access, and rapid elasticity to adjust to changing computing demands. Because medicine is becoming increasingly data-intensive, cloud computing in healthcare research is gaining popularity (Griebel et al. 2015). As a result, health researchers are suggesting cloud computing as a new commercial model for exchanging biomedical data (Rosenthal et al. 2010). Building a stable and sustainable data-sharing situation has garnered fast-expanding attention in both academic research and the healthcare business, thanks to the widespread use of healthcare information and communication technologies. Healthcare cloud computing addresses the requirement for healthcare information to be shared directly with multiple healthcare providers over the internet, regardless of their location or data volume. The cloud

U. A. Okengwu (✉)
Department of Computer Science, University of Port Harcourt, Port Harcourt, Rivers State, Nigeria
e-mail: Ugochi.okengwu@uniport.edu.ng

E. O. Nwaichi (ed.), *Modernity in Health and Disease Diagnosis: The Account from STEM Women*, Sustainable Development Goals Series, https://doi.org/10.1007/978-3-031-34963-8_2

computing paradigm has led to collaborative healthcare research, which provides a cost-effective, adaptable, secure, and privacy-protected cloud-based platform for healthcare researchers throughout the globe to readily explore heterogeneous medical data for health-related analysis (Mu-Hsing et al. 2012). The global healthcare cloud computing industry is estimated to reach $35 billion by 2022, according to BCC research, with an annualized growth rate of 11.6 percent (BCC Research 2022).

2.2 Cloud and Computing in the Cloud

The term "cloud" refers to internet-accessible servers, as well as the software and databases that run on such servers. Cloud servers can be found in data centers all around the world. Users and businesses who use cloud computing do not have to manage physical servers or run software programs on their personal computers (Seth and Singh 2017). "It is a version for allowing convenient, on-demand network get admission to a shared pool of configurable computing sources (example: networks, servers, storage, programs, and services) that may be hastily provisioned and launched with minimal control effort or cloud provider interaction," according to the National Institute of Standards and Technology (NIST) (Badger et al. 2011).

2.3 Cloud Computing's History

The idea of cloud computing started in 1950, with the deployment of centralized servers accessible via static clients. Since then, cloud computing has progressed from static clients to dynamic clients, and from software to services, with a focus on providing a common virtual platform for academic research and thereby bridging the distance gap. In the mid-1960s, computer researcher John McCarthy proposed the concept of timesharing, allowing a corporation to use a pricey mainframe at the same time. This computing is a pioneer of cloud computing and has made a significant impact on the development of the internet. In 1969 J.C.R. Licklider proposed the idea of an "Intergalactic Computer Network" or "Galactic Network" (a computer networking concept comparable to today's net) in order to improve ARPANET (Advanced Research Projects Agency Network); his vision was for everyone on the planet to be connected, with access to packages and information available at any time and from any location (Herring 2015). In 1970, virtualization tools such as VMware made it possible to operate numerous operating systems in a distant environment. It became possible to run a whole new computer (virtual machine) within another operating system.

Prof. Ramnath Chellappa in Dallas in 1997 coined the phrase "Cloud Computing," which he defined as "a computing paradigm where the boundaries of computing can be decided by monetary cause rather than technological restrictions alone" (Mishra 2014). In 1999, Com pioneered the idea of transferring corporate programs via a simple website. The services firm outlined how professional and mainstream software companies may deliver apps through the internet. The first public version of Xen, which generates a Virtual Machine Monitor (VMM), also known as a hypervisor, a software device that allows the execution of several digital visitor running systems on a single system, was released in 2003 (Barham et al. 2003). In 2006, Amazon expanded its cloud services, first as the Elastic Compute Cloud (EC2), which allowed users to connect to computers and run their own programs on them, all while being hosted in the cloud. Then they removed Simple Storage Service from the eq. (S3). Customers and the industry received the pay-as-you-go version as a result, and it has now become normal practice. The use of cloud computing in health-related research is increasing due to the increase in data centers around the world.

2.3.1 Cloud Computing Infrastructure

The hardware and software program elements required to set up cloud computing infrastructure are referred to as cloud computing infrastructure. Computing power, networking, servers, storage,

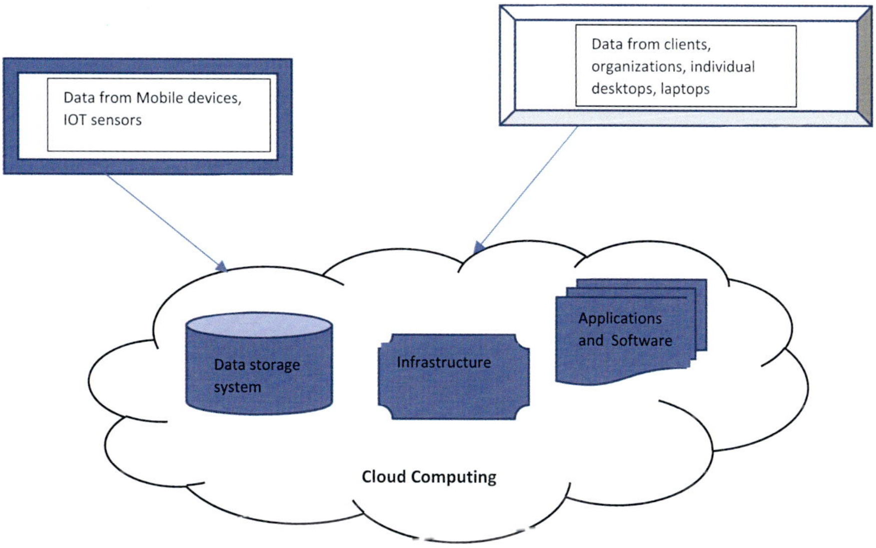

Fig. 2.1 Conceptual representation of cloud computing

virtualization software, services, and control equipment are all included in a cloud computing version's computing requirement. Servers, network switches, memory, and storage clusters are all represented as virtual sources in the virtual infrastructure. Cloud infrastructure has the same capabilities as physical infrastructure, but with the added benefit of cheaper total cost of ownership, better flexibility, and scalability. An abstraction layer, which virtualizes and logically delivers assets and services to users via application interfaces and API-enabled command-line or graphical interfaces, is also included in cloud infrastructure as seen in Fig. 2.1. These virtualized assets are hosted by a provider issuer or IT branch and given to clients via a network or the internet in cloud computing. Virtual machines and additives, such as servers, memory, network switches, firewalls, load balancers, and storage, are among the sources.

2.3.2 Cloud Infrastructure Components

(i) Hypervisor: Hypervisor is a low-level software or firmware. It performs the function of a virtual device manager. It allows several customers to share a physical instance of cloud assets.

(ii) Software for management: Management software aids in the maintenance and configuration of infrastructure.

(iii) Software for deployment: Installing and combining the utility in the cloud is made easier using deployment software.

(iv) Network: The cloud infrastructure's most essential concern is the network. It enables cloud services to be connected through the internet. The customer can design the network path and protocol, making the network as a service delivery over the internet possible.

(v) Server: The server assists in resource-sharing computations and provides a variety of services such as useful resource allocation and reallocations, resource tracking, and security, among others.

(vi) Storage: Many copies of storage are kept on the cloud. If one of the resources fails, it extracts any other resource using those copies. The infrastructure's primary goal is to aid scalability.

2.3.3 Cloud Computing Basic Concepts

As shown in Fig. 2.2, there are certain services and concepts behind the scenes that make cloud

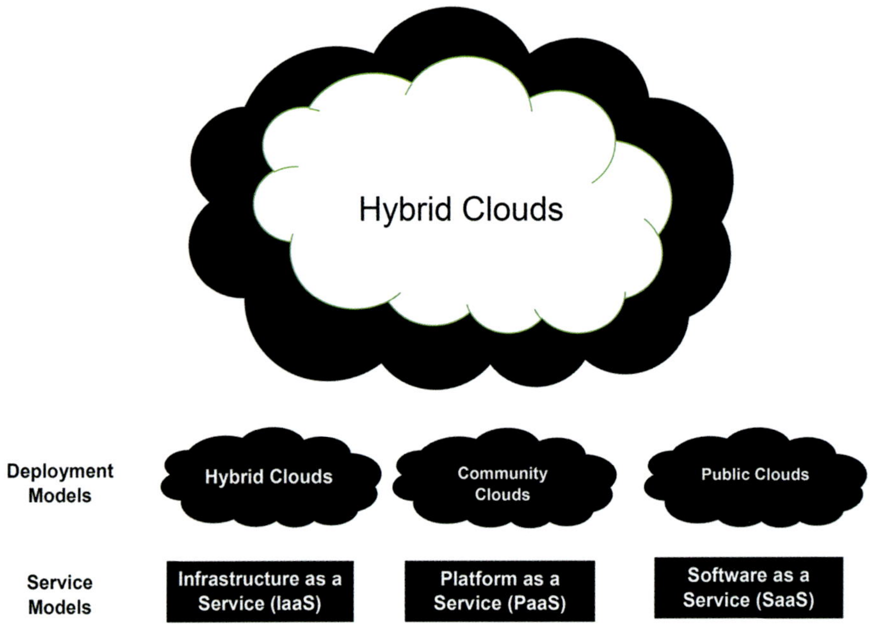

Fig. 2.2 Basic concepts of cloud computing. Source: https://innovationsoncloud.wordpress.com/2012/08/10/systemsconcepts-similar-to-cloud-computing/ (Innovations on Cloud 2012)

computing possible and useful for end users (McAfee 2011).

(i) Models of deployment.
(ii) Service models.

2.4 Models of Cloud Computing Deployment

The type of access to the cloud and how the cloud is positioned are defined by deployment models. Figure 2.3 displays the four sorts of cloud models: public, private, hybrid, and community.

2.4.1 Public Cloud

A public cloud is a sort of cloud computing in which a service provider makes computing resources accessible to the general public via the internet. When we say "cloud" in casual conversation, we usually mean public cloud. Virtualization and today's network era are used in public cloud computing products to provide on-demand scalable computing and storage. Many of the world's most important infrastructure customers, as well as a large number of smaller firms, rely on cloud infrastructure. Public clouds are provided by businesses that often provide a wide range of cloud-related services and resources and charge customers for their use. Even though they are referred to as public, this refers to the fact that anyone can utilize them. It does not imply that any software or records you install are accessible to the public. You have complete control over who has access to any software or data you deploy on public clouds. Amazon EC2, Microsoft Azure, Google Cloud, DigitalOcean, FlexiScale, and Rackspace are examples of public clouds. The fundamental advantage of these platforms is that sources are often available on-demand, with only a brief time between a request for a resource and its availability to the consumer. In comparison to traditional remotely hosted servers, which may have a minimum charge time of a month or longer, public clouds may be far less secure due to their openness.

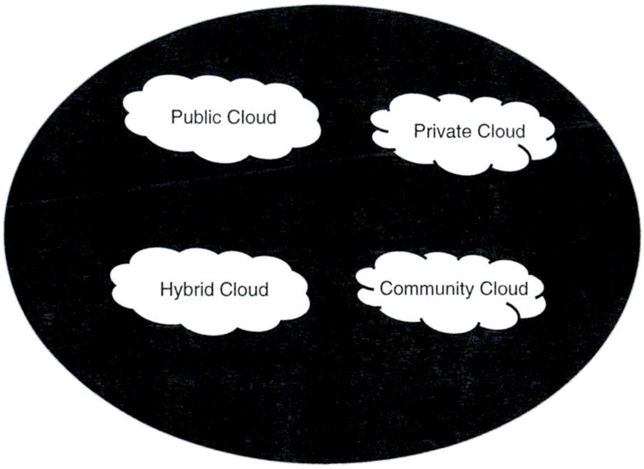

Fig. 2.3 Cloud deployment models. Source: https://www.tutorialspoint.com/cloud_computing/cloud_computing_tutorial.pdf (Tutorialspoint n.d.)

2.4.2 Private Cloud

Private Cloud refers to cloud computing services delivered over the internet or a private internal network to a select group of users rather than the general public. IT provider suppliers began offering private clouds that have the same benefits as the public cloud in response to the requests of enterprise infrastructure hosting clients for privacy and control. The key contrast is that all cloud servers are owned and controlled by a single company. Custom private clouds enable businesses to make use of the benefits of virtualization in a completely secure and private environment that can be tailored to the specific needs of their workloads. Private clouds are set up within a specific private community and allow systems and services to be provided within an enterprise. This could be a single group or a group of people working together across different places. The organization, or "cloud proprietor," now has control over the cloud implementation. Personal clouds are frequently used for mission-critical systems within businesses, as well as for scenarios involving programs or data that cannot be hosted on third-party clouds (e.g., medical applications with sensitive records). Personal clouds can also use any choice of computing gear that is deployed with appropriate middleware software, which could involve reusing existing hardware. The private cloud is managed by a single company. Because of its intimate nature, it is more secure.

2.4.3 Hybrid Cloud

Hybrid cloud computing is a type of cloud computing that combines public and private cloud computing to provide computer services. Essential computing assets are handled in private cloud computing, whereas non-critical computing assets are handled in public cloud computing in this setting. Pundits pitted public and private clouds against each other in the early days of cloud computing, speculating on which would acquire dominance. In truth, public and private clouds are complementary technologies, and many users of enterprise infrastructure find a place for both. As a result, hybrid clouds have emerged, which mix public and private cloud platforms, as well as metal-committed servers, which continue to play an important role. Combinations of the classes listed above are also possible. Combining an internal non-public cloud with a public or network cloud platform, for example, can create a hybrid cloud. When the internal non-public cloud becomes overburdened, the external cloud can be used to alleviate demand overflow ("cloudbursting") or to archive data.

2.4.4 Community Cloud

This type of cloud computing is for a shared cloud computing service environment that is targeted at a small number of agencies or individuals, such as banks or heads of purchasing and selling organizations. Some organizations or agencies can use the community cloud to access systems and services. It distributes infrastructure among a group of enterprises in a network. It is designed for certain groups and can be handled internally by organizations or by a third party. These could be organizations that have formed strategic alliances or provide services to specific groups. These clouds may be more expensive to provision than public clouds, especially if existing hardware is to be used (due to the possibility of an initial set-up overhead), but they may also provide services suited to the community's needs.

It can be installed using the same frameworks as private clouds, but it may require additional higher-level services to connect users across several sites. NASA's Nebula is an example of a previous community cloud, while EGI's Federated Cloud is another. However, many previous community cloud users have switched to hybrid clouds, and the number of community clouds is dwindling.

2.4.5 Models of Cloud Computing Services

Although there are more cloud computing services, Fig. 2.4 shows the three primary service types.

(i) Platform-as-a-Service (PaaS) is a cloud computing model in which a cloud computing service provider provides end users or clients with hardware and software. Google App Engine and Microsoft Azure are two examples. The cloud computing subscriber may be responsible for software services, while the cloud computing provider may be responsible for runtimes, security and integration, databases and servers, virtualization, server hardware, storage, and networking. With PaaS, the service provider provides clients with a platform on which they can host, execute, manipulate, or expand packages without having to build or maintain their own on-premises or cloud infrastructure. Customers gain access to cloud resources via a platform interface at a higher level. PaaS solutions provide services that simplify obligations that an infrastructure-as-a-service (IaaS) user should have handled on their own. PaaS encompasses a wide range of services, ranging from popular ones built on top of infrastructure clouds to much improved domain-specific architectures. Virtualized hardware is typically used to offer resources, and the user is responsible for controlling the resources and installing their applications onto them. This provides organizations with access to the whole IT infrastructure they require, whether temporarily or permanently. Companies benefit from the low cost of this service model because they don't

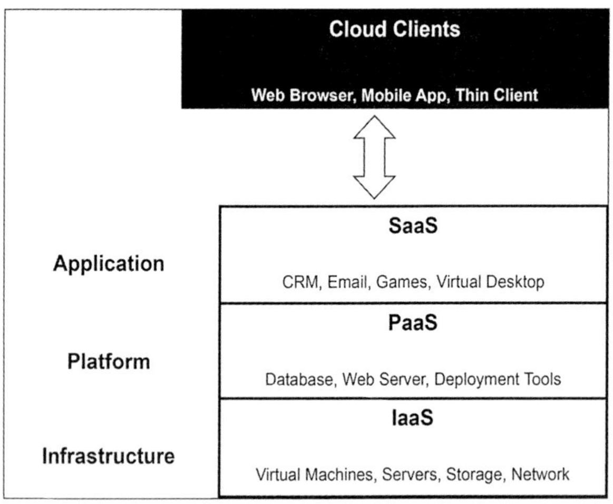

Fig. 2.4 Three common service models. Source: https://www.tutorialspoint.com/cloud_computing/cloud_computing_tutorial.pdf (Tutorialspoint n.d.)

have to invest in gear that will only be utilized a few times.

(ii) Infrastructure-as-a-Service (IaaS) is a provider model in which infrastructure is hosted in the cloud, both private and public, rather than in a traditional on-premises data center. Cloud infrastructure products, often known as IaaS, contain automated and easily expandable computer assets. Amazon Web Services EC2 and S3 are two instances of infrastructure that is made available to consumers on demand while being totally managed by the providing business. This cloud computing subscriber will be responsible for applications, runtimes, security and integration, databases, and servers, while the cloud computing provider will be responsible for virtualization, server hardware, storage, and networking. Customers often get on-demand access to assets and pay only for what they use. Virtualized hardware is often used to offer resources, and the customer is responsible for controlling the resources and installing their programs onto them. This allows groups to gain access to the whole IT infrastructure they desire, whether temporarily or permanently. The low cost of this service version allows agencies to avoid investing in hardware that will only be used for a limited period of time.

(iii) Software-as-a-Service (SaaS) is a model for software distribution in which a cloud provider hosts packages and makes them available to consumers via the internet. In this service version, a software producer may enter into an agreement with a third-party cloud company to host the software or software program, such as Salesforce's customer relationship management software and Google Docs, which are sometimes with giant IT groups like Microsoft, Google, and Amazon as the software vendor. In this model, a business hosts software and applications in the cloud and then sells them to customers on a subscription basis. Adobe, which offers a variety of tools, is an excellent illustration of this. Customers can subscribe to one application, including Adobe Photoshop, or one of the different bundles of packages, depending on their preferences. Customers no longer have to deal with the complexity of gaining access to and managing their own resources, as well as implementing and managing their software applications on these resources. The end user is unaware of the underlying infrastructure, and the SaaS provider oversees capability and scaling to ensure that they can handle the growing number of carrier clients.

Software-as-a-service apps include web-based completely email packages and a wide range of dispensing software programs that employ a user on a user's system to connect with a far-flung software provider.

(iv) Anything-as-a-Service (XaaS) is a service that combines network-as-a-service, business-as-a-service, identity-as-a-service, database-as-a-service, and strategy-as-a-service into one package. Despite the fact that the "X" might stand for anything in this circumstance, XaaS is an abbreviation. As a result, XaaS stands for "anything as a service." It's a term that refers to any software, applications, services, games, and other items that are installed on your computer or other device via the cloud rather than on-premises.

Microsoft Azure, Amazon Web Services (AWS), and even Google Apps are among the leading providers of XaaS. Malware-as-a-Service (MaaS), a type of organized cybercrime in which customers enlist in a service that provides them with malware to mount attacks, is a criminal version of this sort of service.

2.5 Cloud Computing Benefits

(i) Cost Effectiveness: The cost of cloud computing, particularly public cloud, is relatively low since it functions at a high efficiency with optimal usage. It only requires an internet connection, which results in a reduction in the operation charge based on the scale of the rise in products and services provided by the research team, as cloud computing is not dynamic.

(ii) Accessibility: It is simple to obtain items such as utilities via the internet. This allows for online program editing and setup at any moment.

(iii) Reliability: It makes use of a wide range of resources from various locations. If one of the resources fails, the public cloud can replace it with another. Through the PaaS concept, it also provides online development and deployment capabilities, as well as a programming runtime environment.

(iv) Flexibility: Public cloud and private cloud can be seamlessly combined, providing clients with a versatile approach.

(v) Location Independence: Cloud services are delivered through the internet, allowing for location independence.

(vi) Utility Style Cost: It is entirely based on a pay-per-use approach, with resources available whenever the consumer requires them.

(vii) Excessive Scalability: Cloud assets can be scaled up or down to meet the needs of the user.

(viii) Self-Service: Assets in cloud computing can be used without having to communicate with the cloud service provider. One of the most important features of cloud computing is its ability to be accessible from anywhere with a reliable internet connection.

2.5.1 Health-Related Academic Research Benefits of Cloud Computing

(i) Improve the Research Quality.
With medical data, cloud storage for healthcare makes collaborative research a breeze. This results in the digitization of healthcare data via cloud-based data storage. A cloud should give you access to resources that are faster, better, or more scalable, allowing you to do work that you can't complete with your current resources, whether they are within your department, institution, or the broader academic community (Pranay et al. 2013). Similarly, cloud resources may enable you to obtain more accurate findings than you can currently.

(ii) Increase the Amount of Research Done.
By offering a platform for Big Data analysis and the development of predictive models for epidemic outbreaks, cloud computing for healthcare can make Big Data healthcare research simple and possible. Clouds can help you increase the amount of research you can do in a given amount of time. This can be accomplished by gaining access to more efficient assets that will allow you to complete your analyses more quickly. Similarly, a cloud can provide a way to supplement your existing assets in order to meet peaks in your study.

(iii) Improve the Cost-Effectiveness of the Project.
In contrast to on-site storage infrastructure, cloud computing in healthcare research will cut data storage costs. Cloud computing has the potential to improve your cost-effectiveness in a variety of ways, allowing you to get the most out of your budget. It may be less expensive to use the cloud than to order, set up, and manage new gear within your organization. Because of the pay-per-use model, it offers a significant reduction in the expense of conducting academic research.

(iv) Minimize the Influence on the Environment.
Cloud computing has the ability to reduce your environmental footprint. Virtualized physical resources in the cloud can host a variety of independent digital hardware that can be used for various tasks or establishments to optimize available CPU capability (Waga et al. 2014). If businesses had distinct physical hardware resources, they may remain inactive for long periods of time, wasting electricity and space and necessitating cooling. Public cloud structures, in which resources are shared and built for usage by a wide number of users with vary-

ing needs, may also allow for the reuse, or recycling, of existing technology, which can help to reduce environmental impact.

(v) Makes Healthcare Research More Secure and Private.

Using HIPPA (Health Insurance Portability and Accountability Act of 1996)-compliant cloud storage services, healthcare cloud computing provides sophisticated and improved data protection.

2.5.2 Cloud Computing Contribution to the Achievement of Sustainable Development Goal 3

The benefits of cloud computing have exploded in recent years thanks to the use of collaboration platforms in the digital world, and this has had a big impact on human health. The self-service internet infrastructure known as cloud computing, which allows users to access computing resources from any location at any time, has demonstrated a strong relevance in advancing the objectives and key performance indicators of Sustainable Development Goal 3: Healthy Lives and Well-Being. Cloud computing will facilitate the accomplishment of universal health coverage by making cloud-based health resources widely available and affordable to most low-income nations through its collaborative research platform. This will improve the achievement of SDG 3 targets 3.6 and 3.8. However, the research and development of vaccines and medications for developing countries will be completed quickly because cloud computing will encourage quick, on-time, and real-time communication among researchers working in these fields. This will invariably result in the availability of these medications and vaccines, particularly for developing nations.

Information Technology (IT) firms like NetApp, athenahealth, Microsoft Azure, AllScripts, CareCloud, Medable, ClearDATA, Medsphere Systems, and other firms like Pfizer have taken advantage of cloud-based frameworks to offer low-cost healthcare resources, including cloud computing automation resources, real-time clinical data delivery, HIPAA implementation, and improvements to the manufacturing and distribution processes.

2.6 Cloud Service Providers in the Top Ten

The top ten cloud computing providers are: Amazon Web Services (AWS), Microsoft Azure, Google Cloud, Alibaba Cloud, IBM Cloud, Oracle, Salesforce, SAP, Rackspace Cloud, and VMWare.

2.7 Computing Platforms in the Cloud

(i) Engine Yard is a rails application that runs on a cloud computing platform.
(ii) Enomaly: The IaaS platform is presented.
(iii) FlexiScale: This company provides a cloud computing platform that enables flexible, scalable, and automated cloud architecture.
(iv) GCloud3: GCloud3 is a platform that enables non-public cloud solutions.
(v) Gizmox: This is well-suited for creating new net apps as well as modernizing legacy apps based on ASP.Internet, DHTML, and other technologies.
(vi) GoGrid: Provides consumers with the ability to install internet and database cloud services.
(vii) LongJump: This is a PaaS that provides a business utility platform.
(viii) Microsoft: Windows Azure is a cloud computing platform that provides an environment for developing cloud apps and services.
(ix) OrangeScape: This provides a PaaS for non-programmers, allowing them to create apps as easily as they would a spreadsheet.
(x) RackSpace: It provides on-demand servers via a cloud-based infrastructure of virtualized servers.

(xi) Amazon EC2 (Elastic Compute Cloud): This allows users to configure and modify computational resources while operating them on Amazon's infrastructure.
(xii) Salesforce.Com: This site has a simple interface that allows users to log in, create an app, and publish it to the cloud.
(xiii) Appistry: The CloudIQ platform from Appistry excels at providing runtime applications. This framework is extremely useful for developing scalable, service-oriented packages.
(xiv) AppScale: AppScale is a free open-source platform for Google's App Engine.
(xv) AT&T: This allows users to access digital servers and controls the virtualization infrastructure, which includes the network, server, and storage.

2.8 Conclusion

The cloud has many potential benefits for healthcare researchers, but whether or not these benefits materialize depends on a variety of factors, including the cloud platform used; the types of assets employed; the nature of the application; and the time, money, and effort available. Similarly, each benefit may necessitate trade-offs. For example, you may gain the ability to perform a far greater number of jobs in parallel, but at the cost of a slight increase in execution time in line with activity due to various resource specifications; you may have access to more processing capacity at the cost of not purchasing the underlying gear; or you could avoid purchasing and installing your own cluster, lowering your capital expenditure but increasing your operational expenditure over time. The practice of installing remote servers accessed via the internet to store, manage, and process healthcare-related data is known as cloud computing in healthcare. In contrast, setting up an on-site data center with servers or hosting the data on a personal computer are both options. Many pharmacology vendors are also providing PaaS and SaaS solutions to help with research and medication development. Cloud computing is becoming a more important part of healthcare Research and Development as a result of the explosion of data from next-generation sequencing and the growing role of biologics in the research process.

References

BCC Research. Healthcare cloud computing technologies report. 2022. https://www.bccresearch.com/market-research/healthcare/healthcare-cloud-computing-technologies-report.html.

Badger L, Grance T, Patt Corner R, Voas J. Cloud computing synopsis and recommendations. *Recommendations of the National Institute of Standards and Technology*, C O M P, 2011 84.

Barham P, Dragovic B, Fraser K, Hand S, Harris T, Alex H, Neugebauer R, Pratt I, Warfield A. Xen and the art of virtualization. *Proceedings of the Nineteenth ACM Symposium on Operating Systems Principles – SOSP* 2003 37. 164–177. DOI: https://doi.org/10.1145/945445.945462.

Colin TSX, Felicia TWX. Benefits and challenges of the adoption of cloud computing in business. Int J Cloud Comput Serv Archit (IJCCSA). 2016;6(6):1–8.

Griebel L, Prokosch HU, Köpcke F, Toddenroth D, Christoph J, Leb I, Engel I, Sedlmayr M. A scoping review of cloud computing in healthcare. BMC Med Inform Decis Mak. 2015;15:7. https://doi.org/10.1186/s12911-015-0145-7.

Herring MY. Social media and the good life: do they connect? McFarland; 2015.

Innovations on Cloud. Systems concepts similar to cloud computing. 2012. https://www.innovationsoncloud.wordpress.com/2012/08/10/systemsconcepts-similar-to-cloud-computing.

McAfee A. What every CEO needs to know about the cloud, vol. 89. Harvard Business Review; 2011.

Mishra D. Cloud computing: the era of virtual world. Int J Comput Sci Eng. 2014;3:204–9.

Mu-Hsing K, Andre K, Elisabeth B, Feipei L, Sarangerel D, Chinburen J. 'Cloud computing for healthcare research in information sharing'. *Conference proceedings 2012 IEEE 4th international conference on cloud computing technology and science (Cloudcom)* 2012.

Pinal VC. Cloud computing in distributed system. Int J Eng Res Technol (IJERT). 2012;1(10):2–4.

Pranay K, Sumitha K, Uma RN, Jangaon. Effective ways cloud computing can contribute to education success. Adv Comput: An International Journal (ACIJ). 2013;4:4.

Rosenthal A, Mork P, Li MH, Stanford J, Koester D, Reynolds P. Cloud computing: a new business paradigm for biomedical information sharing. J Biomed Info. 2010;42(2):342–53.

Seth S, Singh N. Cloud computing: a perspective on next basic utility in IT world. Int Res J Eng Technol (IRJET). 2017;4:7.

Tutorialspoint. Cloud computing tutorial [PDF]. n.d. https://www.tutorialspoint.com/cloud_computing/cloud_computing_tutorial.pdf.

Waga D, Makori E, Rabah K. Utilization of cloud computing in education and research to the attainment of millennium development goals and vision 2030 in Kenya. Univers J Educ Res. 2014;2(2):193–9. Retrieved November 6, 2021 from https://www.learntechlib.org/p/160632/

Significance of Nutritional Etiquette to Women's Health

3

Ododobari Jike-Wai

3.1 Introduction

Nutrition is the study of food and how the body uses it. It can also be defined as the science that provides explanations of how the body uses food nutrients as nourishment for development, optimal functioning, and performance. Etiquette is all about behaving in a manner that shows you are responsible. Consequently, nutritional etiquette is engaging in responsible eating behaviours that guarantee nourishment, proper development, optimal body functioning, and performance throughout our lifetime. Simply put, it is engaging in eating patterns that ensure living in sound health. Nutrition is the pivot for human existence as food is a basic need of man and man has continued to consume local herbs and vegetables (medicinal plants) to improve health, while bringing down diseases (Nwaichi et al. 2019). The effect of inadequate nourishment (poor nutrition) of women negatively affects their ability to effectively procreate and effectively perform other functions in life. This in turn affects the society in the form of poor work capacity and ill health of their offspring (Galloway et al. 2002; ACC/SCN 2000). Most infant and childhood nutritional challenges that are irreversible have their root cause in poor maternal nutrition (Ahmed et al. 2012). Maternal nutritional status is an echo of maternal food behaviour and health status. Nutritional status is the physiological state of an individual which is decided by the body's utilization of nutrients. Malnutrition occurs when there is a restricted supply or poor utilization of food nutrients by the body. Malnutrition is an indicator of poor nutritional status, which is a reflection of poor food behaviour. For women to function optimally both at home and in society there is a need to imbibe good nutritional etiquette.

Nutritional etiquette is essential to women's health because research is continually linking chronic diseases to poor eating behaviours (Chiuve et al. 2012; Fung et al. 2001; Geng et al. 2021; Liese et al. 2015), and maintaining a healthy eating pattern can reduce the risk of developing chronic diseases (Estruch et al. 2018) in line with SDG3 as well as improve brain and general body functions (Carson et al. 2014), and quality of life (Gilbertson et al. 2020). Hence, this chapter seeks to highlight nutritional etiquette that are significant to women's health.

3.2 Nutritional Etiquette

These are practices that individuals especially women should cultivate in order to live in good health. They include:

O. Jike-Wai (✉)
Department of Food, Nutrition and Home Science, Faculty of Agriculture, University of Port Harcourt, Port Harcourt, Rivers State, Nigeria
e-mail: ododobari.jikewai@uniport.edu.ng

E. O. Nwaichi (ed.), *Modernity in Health and Disease Diagnosis: The Account from STEM Women*, Sustainable Development Goals Series, https://doi.org/10.1007/978-3-031-34963-8_3

3.2.1 Eating Right

Eating right involves a wide array of feeding practices an individual must be engaged in to maintain a good nutritional and health status. This encompasses eating adequate diets at all mealtimes. Meals that contain all key nutrients (carbohydrates, proteins, fats/oil, vitamins, and minerals) in their right amounts, every time you are eating. Choose foods that suit your age, health status, and activity level and consume them in the right quantities at mealtimes. People of different ages require more or less quantity of foods from the different food groups because of the basic nutrients they supply to the body. At a younger age, more energy-giving, bodybuilding, and protective food are required as they are still growing and in their active stage of life. They require adequate quantities of carbohydrates, proteins, vitamins, minerals, and fatty foods in their meals (Mitchell 2015). While at an older age, nutrient-rich, low-calorie foods and good hydration are essential. Choosing foods that suit your health status is eating with respect to underlying health conditions suffered by an individual. Certain foods exacerbate specific health conditions while others improve those conditions. It is advisable to adhere to the counsel of your physician/nutritionist/dietician concerning your health status and food choices. An individual who is engaged in high energy–expending activities needs to consume high energy–giving foods since the rate of energy expenditure is proportionate to energy gain, leaving no room for storage of excess and or fatigue if the reverse is the case.

The several facets to eating right involve a conscious effort to change our eating behaviour/pattern:

A. *Food portion sizing is an integral part of eating right*—Portion size refers to the quantity of food you choose to consume at each mealtime. Portion size regulates macronutrient fractions for a well-proportioned meal. The main goal of portion sizes is to prevent over-indulging in a single food group which could affect the nutritional and health status of the individual concerned. Individual portion sizes for a normal healthy person can be determined easily through either of the under-listed methods:
   (i) *Using your plate as a portion guide*. This involves dividing your dinner plate into four equal parts with an imaginary line, with half the plate for vegetables and fruit foods, a quarter for carbohydrate foods, and the last quarter for protein foods, while fat and oils are used sparingly as dressing (Brown 2018). An illustration of how a plate can be used as a portion guide is shown in Fig. 3.1.

   The use of smaller dinner wares (plates, spoons, and cups) is another method of controlling the quantity of food consumed. It is effective as their sizes influence the quantity of their content (served food) and the amount con-

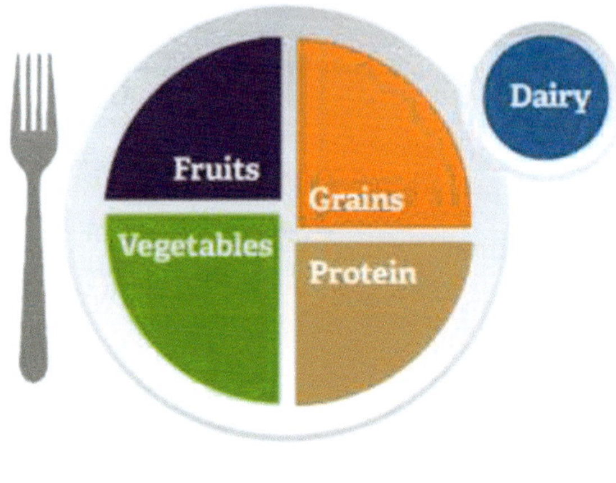

Fig. 3.1 How a plate can be used as a portion guide. Source: https://www.amazon.com/dp/B07C2FFM3M?ots=1&linkCode=ogi&tag=bp_links-20&ascsubtag=[artid|2089.g.22825061[src|[ch|[lt

sumed. Studies have shown that eating with larger dinner wares results in unintentional overeating (Wansink et al. 2005), further asserting the important role visual cues play in curbing overeating (Scheibehenne et al. 2010; van Kleef et al. 2012). Exchanging your larger dinner plates for smaller ones reduces the helping of food and prevents overeating, as you feel just as full eating from a smaller plate as from a larger one (Vartanian et al. 2008). Alternatively, commercially sold portion control plates can be used for food portion control. Its principle is similar to dividing your dinner plate into quartered sections for different food groups. Commercially sold portion control plates are ready-to-use plates that have already been partitioned into sections by the manufacturer. They are available in stores in different designs as can be seen in Figs. 3.2 and 3.3.

Worthy of note is the fact that these portion control measures are a rough guide, as portion sizes are to regulate macronutrient consumption for a healthy meal. People have different dietary needs, which vary according to physical activity level, age, gender, and health status.

(ii) *Using your palm as a portion guide.* Here your palm size is for protein foods, your cupped palm is for carbohydrate foods, your fist size is for vegetables and fruits, while your thumb size is for fat and oil (Brown 2018). The illustration is shown in Fig. 3.4.

B. *Prioritize seasonal food*—This is because nutrient density is highest when the food is in the season than when it is out of season or processed and on the shelf. Seasonal produce is usually harvested at its peak so it retains its full nutrient and vitamin content (Morrison 2019), because it has naturally ripened on the vine or matured in the ground and will have a more complex and rich flavour that appeals more to the taste. However, by eating foods in season you minimize your intake of chemical preservatives and pesticides that are used to preserve these foods when they are off-season, some of which are detrimental to health in the long run.

C. *Use nutrient-conserving food preparation methods*—This helps retain most of the nutrients in the food, as no single food item contains one nutrient alone. A review by Carmody and Wrangham (2009) validates that cooking food improves digestion and raises the absorption of many nutrients. According to *New York Times* (July 7, The New York Times Archives 1982), 'the three R's for nutrient preservation are to reduce the amount of water used in cooking, reduce the cooking time and reduce the surface area of the food

Figs. 3.2 and 3.3 Samples of some commercially sold portion plates. Source: Adapted from https://www.amazon.com/dp/B07C2FFM3M?ots=1&linkCode=ogi&tag=bp_links-20&ascsubtag=[artid|2089.g.22825061[src|[ch|[lt|

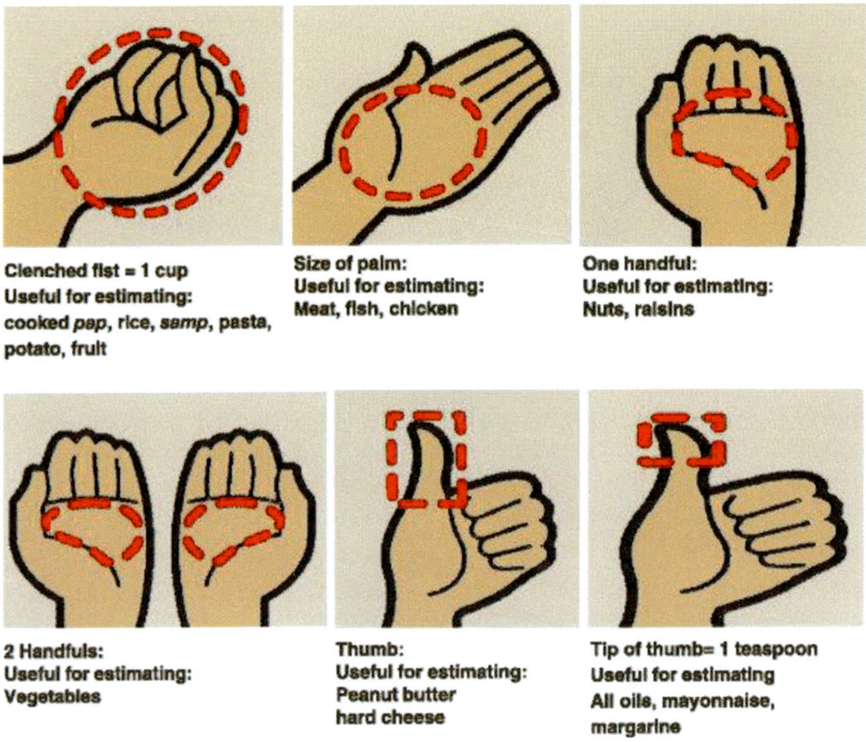

Fig. 3.4 Your palm as a guide to estimate portion size. Source: Adapted from Creative Commons Attribution 4.0 International

that is exposed'. Cooking methods that involve less water, pressure cooking, steaming, stir-frying, and microwaving cause minimum damage to nutrients.

Several vitamins and some minerals (potassium, magnesium, calcium, and sodium) are liable to destruction by air, acid, light, alkali, heat, water, keeping time, and enzyme activity in the foods themselves. However, nutrient losses can be reduced by the precaution taken in choosing, storing, and cooking foods. Cooking methods influence the nutritional content of meals (Spritzler 2019). This is more pronounced with vitamin C and the B-group vitamins which are sensitive to heat and air. Loss of these nutrients increases with longer cooking time and higher temperatures.

D. *Engage in healthy snacking*—This entails eating fruits and vegetables as a snack rather than pastries. If you must eat pastries, eat a smaller portion or as a meal. Healthy snacking has been said to improve overall health, control cravings, combat weight gain, regulate mood, improve brain power, and provide the energy you need to keep going all day (Baertlein 2016). Healthy snacking reduces stress and improves focus and productivity (Sheridan 2020) as SDG3 intended. Studies show that what staffs and students consume during the workday influences mental acuity, energy, and overall performance (4imprint blue papers 2009).

E. *Have a variety of food from the different food groups in a meal*—This ensures that you get a healthy supply of varied and required nutrients in your diet as the monotony of food kills appetite. Consuming foods from different food groups guarantees an adequate supply of essential nutrients needed for optimum health and well-being. The five basic food groups are:

- Bread group: Cereals, rice, pasta, noodles, and other grains.
- Vegetable group: This group is subdivided into two, namely: non-fat vegetables and starchy vegetables.
- Fruit group: Citrus fruits, mangoes, bananas, pawpaw, guava, pineapples, etcetera.
- Milk group: Whole, skimmed, evaporated, condensed milk, yogurt, and cheese.
- Meat group: Beef, fish, poultry, other fleshy foods, and eggs.
- Fat group: This includes oil seeds, margarine, and butter (Okeke et al. 2011).

F. *Limit patronage of food vendors*—Limiting visits to food vendors is important because the kind of ingredients used in meal preparation cannot be ascertained correctly. Food vendors are out to make profits in their businesses and so may imbibe practices that maximize profit yet are unhealthy to their customers. Studies have shown that of the entire menu served by food vendors, only 20% were healthy (Reznar et al. 2019). If you must eat from food vendors, ask for half a portion. In another study (Seguin et al. 2016), increased consumption of food away from home was associated with higher body mass index and lower consumption of fruits and vegetables, both of which have health implications.

G. *Read labels on packaged foods*—The practice of reading labels will help the consumer make an informed choice during purchase. Labels give you information on the nutrient content of the food product, manufacturing and expiration date, ingredients used in production, and some go further to tell you serving size. Reading food labels is an easy way of comparing food products to purchase those with maximum nutritional values (LaBarbera 2012). It also helps individuals make healthy choices about the foods they are buying. On the other hand, you can use food labels to find food products higher in specific nutrients you sort for.

H. *Be consistent in your eating routine and avoid eating beyond 7 pm*—The timing of your meals can affect your body's weight regulation, metabolic regulation, heart health, and sleep cycle (Chaix et al. 2014). If altered, these body processes have detrimental effects on the individual which invariably affects the well-being of the person concerned.

3.3 Periodic Self-Assessment of Nutritional Status Using Anthropometry

Anthropometry is the measurement of body dimensions. It is an easy method of assessing nutritional status but cannot be used alone to decide the nutritional state of a person. Anthropometric measurements are quick tell-tale pointers to the nutritional and health status of an individual. Anthropometric techniques are rapid and repeatable. The required equipment is readily available, inexpensive, and portable (Mitchell 2015). However, a skilled and careful measurer, as well as quality equipment that is calibrated regularly, is essential for attaining accuracy and reliability in anthropometric measurements. Some anthropometric indices that can be self-assessed by individuals include:

A. *Body weight measurement*—It is a composite measure of total body size and provides a rough evaluation of overall muscle mass, bone mass, and fat stores (Wardlaw and Kessel, 2002). It is used in screening for unusual growth, obesity, or undernutrition and is a component in equations that predict energy needs and body composition (Mitchell, 2015). Body weight can be measured using a levelled platform scale with a beam and moveable weights or an electronic scale. The subject wears minimal clothing without shoes and stands with weight evenly distributed on both feet on the scale. The weight is read-off to the nearest 0.1 kg or 100 g. Weighing is not done after a full meal or when the bladder is full. The interpretation of body weight is age-dependent. Excess or lower weight predisposes women to varied negative health outcomes that affect productivity and general well-being.

B. *Height*—The height of an individual is measured using a stadiometer, height meter, or a scale fixed to a vertical wall on a level floor. When measuring the height of a person, the subject stands erect on a flat floor without shoes, with weight equally distributed on both feet and heels together and touching the vertical border height meter. Arms hang freely at the sides of the truck with palms facing the thighs. The head is held erect so that the line of vision is at right angle to the body; the subject takes a deep breath and holds that position while the horizontal headboard is brought down firmly on the top of the head (Okeke et al. 2011). The presence of unusually thick hair should be taken into account. The height is then read-off and recorded to the nearest 0.1 m. Height can be measured in inches (in), feet (ft), centimetres (cm), or metres (m). As with weight measurement, the interpretation of height is age-dependent. Height measurement is a component of the equation for calculating body mass index (BMI).

C. *Body mass index* (BMI) – BMI is a measure of body mass based on weight and height measurements. It is used to identify underweight, overweight, or obese individuals. It highly correlates with adiposity but varies with sex, age, and race. This is because BMI is a reflection of muscle mass, bone mass, and fat deposits. Women whose BMIs are not within the normal range are at health risk to themselves and their offspring. BMI is a function of the weight to height ratio which is used to assess overweight and obesity in adult populations. These ratios are also referred to as obesity indices.

Conversely, the effect of low BMI in adults has been linked with a decline in work output, productivity, and income-generating ability, as well as a compromised ability to respond to stressful conditions (Gibson 2005). Women by being homemakers are engaged in multitasking activities and the resultant effect of low BMI is more severe and pronounced in them than their male counterparts. Low BMI is also known as thinness, underweight, or chronic energy deficiency and has been classified by the WHO into mild, moderate, or severe underweights as shown in Table 3.1.

Body mass index is calculated using weight measurement (kg) divided by height measurement (m^2) and the obtained score is interpreted as shown in Table 3.1.

D. *Waist circumference*—Waist measurement is used to assess abdominal fat deposit, which is an indication of one's predisposition to obesity and metabolic disorders (Ross et al. 2020). It is useful in individuals that are 5 ft. (162.4 cm) taller and have a body mass index of less than 35 kg/m^2 (Okeke et al. 2011). Waist measurement is taken to the nearest centimetre using a non-stretch tape at the smallest area below the rib cage and at the level of the umbilicus around the waist with the subject standing erect. Measurements

Table 3.1 The international classification of adult underweight, overweight, and obesity according to BMI

Classification	BMI (kg/m^2)	
	Principal cut-off points	Additional cut-off points
Underweight		
Severe thinness	<16.00	<16.00
Moderate thinness	16.00–16.99	16.00–16.99
Mild thinness	17.00–18.49	17.00–18.49
Normal range		
	18.50–24.99	18.50–22.99
		23.00–24.99
Overweight		
Pre-obese	25.00–29.99	25.00–27.49
		27.50–29.99
Obese		
Obese class I (moderate)	30.00–34.99	30.00–32.49 32.50–34.99
Obese class II (severe)	35.00–39.99	35.00–37.49 37.50–39.99
Obese class III (very severe)	≥40.00	≥40.00

Source: Adapted from WHO global database on BMI (WHO 2006)

greater than 102 cm in men and 88 cm in women are a pointer to abdominal adiposity (excess abdominal fat), a risk factor for obesity (WHO 2008). It is certain that waist circumference alone can be used to assess abdominal fat.

E. *Hip circumference*—The hip is defined as the largest circumference between the waist and the knee (Mitchell 2003). Hip circumference is measured at the point of the greatest circumference around the hip and recorded to the nearest 0.1 cm. Measurement is taken with the subject standing erect, arms by the side of the body, and feet together on a level surface. The measurement is taken with a flexible non-stretch tape in close contact with the skin but not compressing the soft tissues. Hip circumference is one of the indices needed to determine the waist-hip ratio.

The waist and hip circumference measurements are used to calculate the waist-hip ratio (WHR). WHR distinguishes between android (upper-fat deposits around the waist and abdomen) and gynoid (lower-fat deposits around the hip and buttocks) obesities. WHR is used to evaluate visceral and subcutaneous adiposity. It is important to note that high WHR is strongly connected with an increased risk of developing coronary heart disease, stroke, and type 2 diabetes mellitus in both sexes (Gibson 2005). WHR has been shown to vary with sex, race, age, geographical region, and degree of overweight (Gibson 2005). WHR is calculated and interpreted as shown in Table 3.2:

$$WHR = \frac{Waist\,Circumference(\text{cm})}{Hip\,Circumference(\text{cm})}$$

Table 3.2 Waist to hip ratio reference

Male	Female	Health risk based solely on WHR
0.95 or below	0.80 or below	Low risk
0.96 to 1.0	0.81 to 0.85	Moderate risk
1.0 +	0.85 +	High risk

Source: WHO Expert Consultation (2008)

3.4 Good Hydration

Water is essential for normal body functions. The human body is made up of approximately 75% water. Maintaining good hydration is essential to prevent dehydration which has negative health consequences. Some of these include constipation, fatigue, joint and muscle pains, and headaches. Water cannot be compressed and forms the lubricant found in the knee and other joints. It is the basis for amniotic fluid which serves as a shock absorber for the growing fetus.

The amount of water an individual should consume per day to remain hydrated can be determined by dividing the person's weight (in kg) by 0.033 to get its equivalent in litres. For instance, an individual that weighs 85 kg is expected to drink (80 × 0.033 = 2.64 L) of water daily to maintain good hydration. The simplest way to assess your hydration status is by looking at your urine colour. The lighter and clearer the urine colour the better your hydration status, but the darker the urine colour the more dehydrated the individual.

3.5 Regular Exercises

Although not a nutritional etiquette, regular exercise compliments both the health and nutritional status of an individual. Experts say that regular exercise helps strengthen the digestive tract and keep the gut healthy. Research suggests that regular exercise benefits the digestive system by enhancing the microbiota found in the GI tract and reducing the risk of colon cancer (Monda et al. 2017). Exercising intensifies blood flow towards the muscles and digestive tract, as well as prevents constipation and other illnesses. The WHO has put forth several recommendations with respect to physical activity for all age categories for the maintenance of good health. Adults between the ages 18 and 64 years ought to perform at least 150–300 minutes of moderate-intensity aerobic physical activity; or at least 75–150 minutes of vigorous-intensity aerobic physical activity; or an equivalent combination of moderate- and vigorous-intensity activity throughout the week, for significant health benefits (WHO 2020). On the other hand, it is recommended that pregnant and postpartum women should take part in at least 150 minutes of moderate-intensity aerobic physical activity throughout the week for substantial health benefits (WHO 2020). Walking is the easiest form of exercise. Taking long walks in cool weather also helps keep illnesses away and improve our health. Regular exercise keeps the woman's body fit at all times.

3.6 Conclusion

Women's adherence to nutritional etiquette will boost their work and reproductive capacities as well as prevent malnutrition and its outcomes in their offspring. Good nutritional etiquette guarantees a healthier and productive tomorrow which will be beneficial to mothers, children, the household, and society at large. The application of nutritional etiquette in everyday living requires dedication and discipline on the part of the individual.

References

4imprint blue papers. *What's in the breakroom: Employee Diet and Productivity.* 2009. Retrieved from https://info.4imprint.com/wp-content/uploads/1M-03-1009-Blue-Paper-A-Malnourished-Workforce.pdf

Administrative Committee on Coordination (ACC)/Sub-Committee on Nutrition (SCN) and International Food Policy Research Institute (IFPRI) (February, 2000) Fourth Report on the World Nutrition Situation (Geneva: ACC/SCN, 2000); and Commission on the Nutrition Challenges of the 21st Century, Ending Malnutrition by 2020: An Agenda for Change in the Millennium. Accessed [March 22, 2021] from https://www.ifpri.org/publication/4th-report-world-nutrition-situation

Ahmed T, Hossain M, Sanin KI. Global burden of maternal and child undernutrition and micronutrient deficiencies. Ann Nutr Metab. 2012;61(1):8–17. https://doi.org/10.1159/000345165.

Baertlein L. Healthy snacking benefits. Diet & Nutrition. 2016; Retrieved from https://www.everydayhealth.com/diet-nutrition/meal-planning/healthy-snacking-benefits.aspx

Brown MJ. '9 Tips to measure and control portion sizes' 2018. Retrieved from https://www.healthline.com/nutrition/portion-control.

Carmody RN, Wrangham RW. Cooking and the human commitment to a high-quality diet. Cold Spring Harbor laboratory. 2009;74:427–34. https://doi.org/10.1101/sqb.2009.74.019. PMID, 19843593

Carson TL, Hidalgo B, Ard JD, Affuso O. Dietary interventions and quality of life: a systematic review of the literature. J Nutr Educ Behav. 2014;46(2):90–101. https://doi.org/10.1016/j.jneb.2013.09.005. Epub 2013 Oct 30. PMID: 24183706; PMCID: PMC3982833

Chaix A, Zarrinpar A, Miu P, Panda S. Time-restricted feeding is a preventative and therapeutic intervention against diverse nutritional challenges. Cell Metab. 2014;20(6):991–1005. https://doi.org/10.1016/j.cmet.2014.11.001. PMID, 25470547; PMCID: PMC4255155

Chiuve SE, Fung TT, Rimm EB, Hu FB, McCullough ML, Wang M, Stampfer MJ, Willett WC. Alternative dietary indices both strongly predict risk of chronic disease. J Nutr. 2012;142(6):1009–18. https://doi.org/10.3945/jn.111.157222. PMID, 22513989; PMCID, PMC3738221

Estruch RE, Salas-Salvadó J, Covas MI, Corella D, Arós F, Gómez-Gracia E, Ruiz-Gutiérrez V, Fiol M, Lapetra J, Lamuela-Raventos RM, Serra-Majem L, Pintó X, Basora J, Muñoz MA, Sorlí JV, Martínez JA, Fitó M, Gea A, Hernán MA, Martínez-González MA. PREDIMED Study Investigators. Primary prevention of cardiovascular disease with a Mediterranean diet supplemented with extra-virgin olive oil or nuts. N Engl J Med. 2018;378(25):e34. https://doi.org/10.1056/NEJMoa1800389. PMID: 29897866

Fung TT, Rimm EB, Spiegelman D, Rifai N, Tofler GH, Willett WC, Hu FB. Association between dietary patterns and plasma biomarkers of obesity and cardiovascular disease risk. Am J Clin Nutr. 2001;73(1):61–7. https://doi.org/10.1093/ajcn/73.1.61.

Galloway R, Dusch E, Elder L, Achadi E, Grajeda R, Hurtado E, Favin M, Kanani S, Marsaban J, Meda N, Moore KM, Morison L, Raina N, Rajaratnam J, Rodriquez J, Stephen C. Women's perceptions of iron deficiency and anemia prevention and control in eight developing countries. Soc Sci Med. 2002;55(4):529–44. https://doi.org/10.1016/s0277-9536(01)00185-x.

Geng TT, Jafar TH, Neelakantan N, Yuan JM, van Dam RM, Koh WP. Healthful dietary patterns and risk of end-stage kidney disease: the Singapore Chinese Health Study. Am J Clin Nutr. 2021;113(3):675–83. https://doi.org/10.1093/ajcn/nqaa348. PMID: 33381807; PMCID: PMC7948892

Gibson RS. Principles of nutritional assessment. 2nd ed. New York: Oxford University Press; 2005.

Gilbertson NM, Eichner NZM, Khurshid M, Rexrode EA, Kranz S, Weltman A, Hallowell PT, Malin SK. Impact of pre-operative aerobic exercise on cardiometabolic health and quality of life in patients undergoing bariatric surgery. Front Physiol. 2020;11:1018. https://doi.org/10.3389/fphys.2020.01018.

van Kleef E, Shimizu M, Wansink B. Serving bowl selection biases the amount of food served. J Nutr Educ Behav. 2012;44(1):66–70. https://doi.org/10.1016/j.jneb.2011.03.001.

LaBarbera M. *Reading food labels—how does it help buy healthier foods*. May 14, 2012. Retrieved from http://www.nourishinteractive.com/healthy-living/free-nutrition-articles/161-family-facts-importance-reading-food-labels

Liese AD, Krebs-Smith SM, Subar AF, George SM, Harmon BE, Neuhouser ML, Boushey CJ, Schap TE, Reedy J. The dietary patterns methods project: synthesis of findings across cohorts and relevance to dietary guidance. J Nutr. 2015;145(3):393–402. https://doi.org/10.3945/jn.114.205336.

Mitchell MK. Nutrition across the life span. 2nd ed. New Delhi: Med Tech; 2015.

Monda V, Villano I, Messina A, Valenzano A, Esposito T, Moscatelli F, Viggiano F, Cibelli G, Chieffi S, Monda M, Messina G. Exercise modifies the gut microbiota with positive health effects. Oxidative Med Cell Longev. 2017;3831972. Retrieved from https://doi.org/10.1155/2017/3831972.

Morrison L. The benefits of eating seasonal produce and our favorite spring fruits and vegetables. 2019. Retrieved from https://deliciouslyplated.com/food-articles/the-benefits-of-eating-seasonal-produce/.

Nwaichi EO, Essien EB, Ibe UC. Protective and curative effects of Beta vulgaris on dimethyl 2,2-dichlorovinyl phosphate-exposed albino rats. AAS Open Res. 2019;2:26–40.

Okeke EC, Onyechi UA, Ibeanu VN. *Practice of nutrition: a handbook*, 1[st]ed. Enugu Nigeria: University of Nigeria Press; 2011. p. 246.

Reznar MM, Brennecke K, Eathorne J, Gittelsohn J. A cross-sectional description of mobile food vendors and the foods they serve: potential partners in delivering healthier food-away-from-home choices. BMC Public Health. 2019;19(1):744. Retrieved from. https://doi.org/10.1186/s12889-019-7075-8.

Ross R, Neeland IJ, Yamashita S, et al. Waist circumference as a vital sign in clinical practice: a consensus statement from the IAS and ICCR working group on visceral obesity. Nat Rev Endocrinol. 2020;16:177–89. Retrieved from. https://doi.org/10.1038/s41574-019-0310-7.

Scheibehenne B, Todd PM, Wansink B. Dining in the dark: the importance of visual cues for food consumption and satiety. Appetite. 2010;55(3):710–3. https://doi.org/10.1016/j.appet.2010.08.002.

Seguin RA, Aggarwal A, Vermeylen F, Drewnowski A. Consumption frequency of foods away from home linked with higher body mass index and lower fruit and vegetable intake among adults: a cross-sectional study. J Environ Public Health. 2016;2016:3074241. https://doi.org/10.1155/2016/3074241.

Sheridan N. *The importance of snacking*. February 24, 2020. Retrieved from https://pepandlekker.com/blog-1/2020/2/24/the-importance-of-snackingnbsp

Spritzler F. *How cooking affects the nutrient content of foods*. November 7 2019. Retrieved from https://www.healthline.com/nutrition/cooking-nutrient-content 16/03/2021

The New York Times Archives *Preserving the nutrients of food with proper care*. July 7, 1982 Retrieved [16 March 2021] from https://www.Nytimes.Com/1982/07/07/Garden/Preserving-The-Nutrients-Of-Food-With-Proper-Care.Html

Vartanian LR, Herman CP, Wansink B. Are we aware of the external factors that influence our food intake? Health Psychol. 2008;27(5):533–8. https://doi.org/10.1037/0278-6133.27.5.533.

Wansink B, Painter JE, North J. Bottomless bowls: why visual cues of portion size may influence intake. Obes Res. 2005;13(1):93–100. https://doi.org/10.1038/oby.2005.12.

Wardlaw G M. and Kessel M W. *Perspectives in Nutrition*. 5th ed. New York: McGraw-Hill; 2002.

World Health Organization (2006). *Global database on Body Mass Index*. Retrieved [March 23, 2021] from http://www.assessmentpsychology.com/icbmi.htm

World Health Organization. *Waist circumference and waist – hip – ratio*. Report of WHO expert consulta tion. Geneva: World Health Organization. 2008.

World Health Organization. Guidelines on physical activity and sedentary behavior. Geneva: World Health Organization; 2020.

Women at Greater Risk of Alzheimer's: Way Forward

4

Chinagorom Petrolina Ibeachu

4.1 Introduction

Dementia is an umbrella term that houses a group of diseases and illnesses that affect thinking, memory, reasoning, personality, mood, and behavior (Fig. 4.1). Dementia which was previously seen as a Western world problem has gradually become a serious stigma for people in the low- and middle-income countries. It is the most dreaded condition of our time but is highly neglected in Nigeria. The World Health Organization (WHO, 2021), estimates that the number of individuals with dementia worldwide is approximately 55 million, with this number expected to reach approximately 78 million by 2030 and 139 million by 2050.

Alzheimer's disease (AD) is the most common form of dementia, accounting for approximately 60–70% of cases (Villemagne et al. 2013). Alzheimer's disease is a type of brain disease, just as coronary artery disease is a type of heart disease. It is a progressive disease, meaning that it becomes worse with time. It is degenerative, which means cells degenerate or break down. It is irreversible, which means damage cannot be repaired.

Alzheimer's starts with changes in the brain that are unnoticeable to the person affected. Only after years of brain changes do individuals experience noticeable symptoms such as memory loss and language problems. Symptoms occur because nerve cells (neurons) in parts of the brain involved in thinking, learning, and memory (cognitive function) have been damaged or destroyed (Fig. 4.2). As the disease progresses, neurons in other parts of the brain are damaged or destroyed as well (Figs. 4.3 and 4.4). Eventually, neurons in parts of the brain that enable a person to carry out basic bodily functions, such as walking and swallowing, are affected. Individuals become bed-bound and require around-the-clock care. Alzheimer's disease is ultimately fatal (Jack et al. 2009; Braak et al. 2011; Villemagne et al. 2013; Bateman et al. 2012; Reiman et al. 2012; Gordon et al. 2018).

C. P. Ibeachu (✉)
Department of Anatomy, Faculty of Basic Medical Sciences, University of Port-Harcourt, Port Harcourt, Nigeria
e-mail: chinagorom.ibeachu@uniport.edu.ng

E. O. Nwaichi (ed.), *Modernity in Health and Disease Diagnosis: The Account from STEM Women*, Sustainable Development Goals Series, https://doi.org/10.1007/978-3-031-34963-8_4

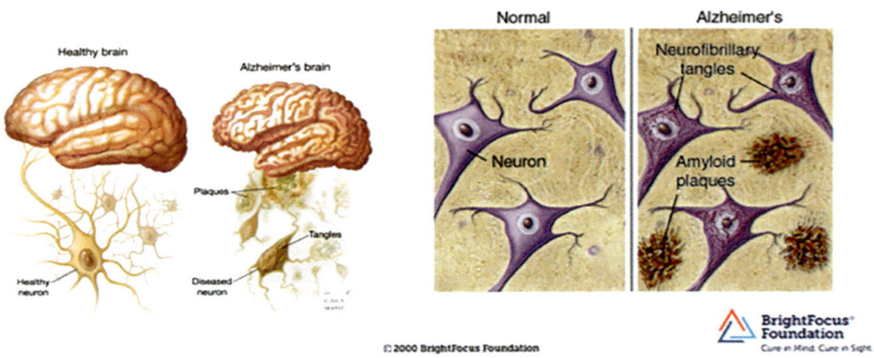

Fig. 4.1 Alzheimer's Association's ten warning signs of Alzheimer's disease

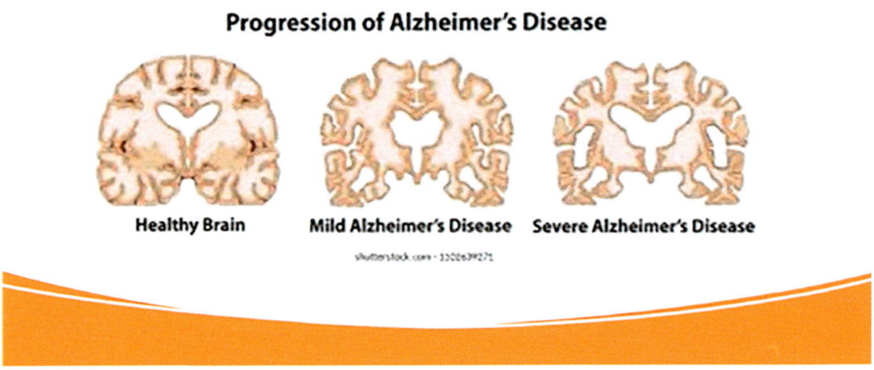

Fig. 4.2 Differences in the brain integrity of normal and Alzheimer's disease conditions

Fig. 4.3 Progression of Alzheimer's disease

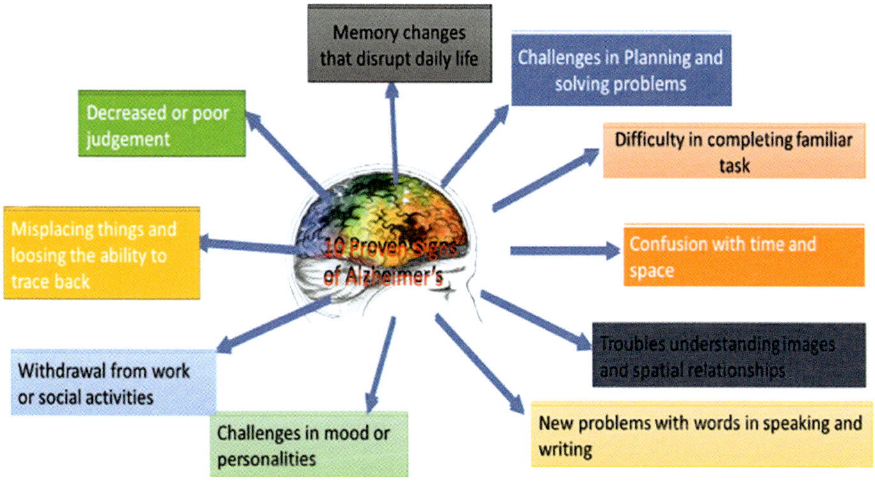

Fig. 4.4 The stages of Alzheimer's disease

4.2 Alzheimer's Disproportionately Affects Women than Men

Clinical studies indicate that Alzheimer's disease disproportionately affects women in both disease prevalence and rate of symptom progression, but the mechanisms underlying this sexual divergence are unknown (Daniel et al. 2018). AD is one of the leading causes of death in the world, currently ranked sixth. Approximately 5.7 million Americans suffer from Alzheimer's, of which almost two-thirds are women (Alzheimer's Association). Of the 6.5 million people aged 65 and older with Alzheimer's in the United States, four million are women and 2.5 million are men (Rajan et al. 2021, Bureau). This represents 12% of women and 9% of men aged 65 and older in the United States (U.S. Census Bureau 2014). Women live longer than men on average, and older age is the greatest risk factor for Alzheimer's (Chene et al. 2015; Seshadri et al. 1997; Hebert et al. 2001). This survival difference contributes to a higher prevalence of Alzheimer's and other dementias in women compared with men.

Every sixty-five seconds, another person develops Alzheimer's disease, and of these newcomers, roughly two-thirds will be women. For a woman over 60, the risk of developing Alzheimer's is twice that of developing breast cancer (Shriver). One study followed 16,926 people in Sweden and found that, beginning around age 80, women were more likely to be diagnosed with Alzheimer's disease than men of the same age (Beam et al. 2018). Similarly, a study based in Taiwan found that one's chances of developing Alzheimer's disease over seven years were greater in women compared to men (Liu et al. 2019). A meta-analysis examining the incidence of Alzheimer's disease in Europe found that approximately 13 women out of 1000 developed Alzheimer's each year, compared to only seven men (Niu et al. 2017). Our study on Sexual Dimorphism in Dementia and Alzheimer Diseases analyzed 82 Alzheimer's and dementia patients admitted in the Psychiatric hospital rumuigbo, Port-Harcourt, Nigeria showed that a greater number of women than men were affected by Alzheimer's disease (Ibeachu et al. 2022) (Fig. 4.5).

Data from the Framingham Study, which enrolled a total of 2611 cognitively intact participants (1550 women and 1061 men) and followed up on many for 20 years, indicated that for a 65-year-old man, the remaining lifetime risk of AD was 6.3% (95% confidence interval [CI], 3.9 to 8.7) and remaining lifetime risk of developing any dementing illness was 10.9% (95% CI, 8.0 to 13.8); corresponding risks for a 65-year-old woman were 12% (95% CI, 9.2 to 14.8) and 19%

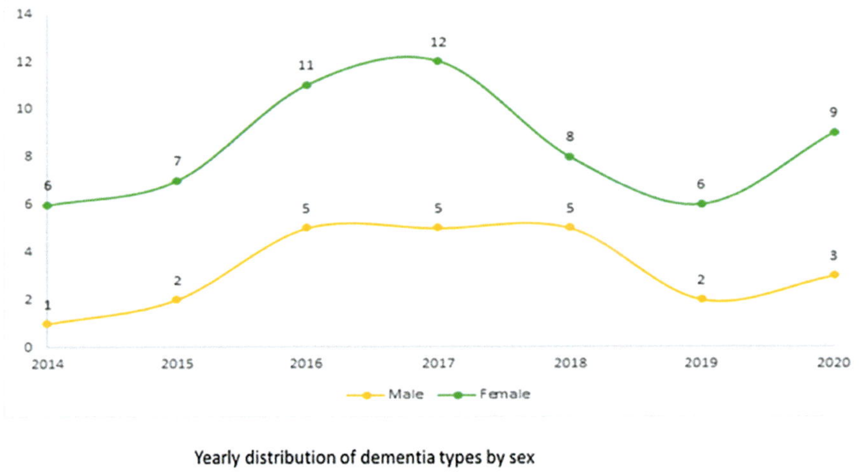

Fig. 4.5 Sexual dimorphism in dementia and Alzheimer diseases: A neuropsychiatric hospital-based study (Ibeachu et al. 2022)

(95% CI, 17.2 to 22.5), almost twice that of men (Seshadri et al. 1997).

Several epidemiologic studies show that neurodegeneration and clinical symptoms occur more rapidly for females once a diagnosis is suspected (Lin et al. 2015; Sinforiani et al. 2010; Hebert et al. 2013). Researchers have hypothesized that this is due to longer female life expectancy or sociocultural detection bias (Mielke et al. 2014); however, there is support that faster progression is due to neurobiological vulnerability in postmenopausal females (Seshadri et al. 1997; Lin et al. 2015). Though the progression of the disease may be more rapid among elderly women, studies conducted in the United States and the United Kingdom suggest that males with AD have a shorter survival time (Kua et al. 2014; Burns et al. 1991). Women are often diagnosed earlier during illness than men, which could confound the determination of postdiagnosis longevity (Podcasy and Epperson 2016).

4.3 Brief History of Alzheimer's Disease

The brain disease that has come to be known as Alzheimer's disease was first described in November of 1901. The chain of events began when Karl Deter brought his wife, Auguste, to the StädtischeHeilanstaltfürIrre und Epileptische (City Hospital for the Mentally Ill and Epileptics) in Frankfurt, Germany. Auguste's behavior made it almost impossible for Karl to work anymore. She would sometimes wake up in the night and scream for hours. Knowing what we know now of Alzheimer's disease, she undoubtedly needed a lot of personal care. Upon her admission, she was interviewed by a resident at the hospital, a young doctor named Aloysius Alzheimer. He had joined the staff at the hospital 13 years earlier and was pursuing his interests in psychiatry and neuropathology. Auguste Deter had problems with memory and with reading and writing, and she showed signs of disorientation. She was diagnosed with presenile dementia.

4.4 The Hallmarks of Alzheimer's

- Amyloid plaques and neurofibrillary tangles are considered the "hallmark" of Alzheimer's disease.
- They form in and around the existing healthy brain cells, often choking them out and causing the brain cells to shrink and die off.
- The brain can shrink up to two-thirds of its original size and weight throughout the course of the disease.

4.5 Risk Factors for Alzheimer's Disease

4.5.1 Non-modifiable Risk Factors

Age Age is the greatest of these three risk factors. The percentage of people with Alzheimer's dementia increases dramatically with age: 5.0% of people aged 65 to 74, 13.1% of people aged 75 to 84, and 33.2% of people aged 85 or older have Alzheimer's dementia.

Family History Those who have a parent, sibling, or child with Alzheimer's are more likely to get Alzheimer's.

Genetics (Heredity) Less than 1% of Alzheimer's cases are caused by deterministic genes (genes that cause a disease, rather than increase the risk of developing a disease). APOE e4- increases the risk of Alzheimer's.

Gender Women are more likely to develop AD than men.

4.5.2 Modifiable Risk Factors

Mild cognitive impairment, high blood pressure, cholesterol, poorly controlled diabetes, limited education, traumatic brain injury, depression, sleep deprivation/insomnia, obesity BMI > 30, alcohol>21/week, smoking, and social isolation.

4.6 Evidence-Based Study on the Sex Differences in Alzheimer's Disease

4.7 Facts about Women

1. Stress-related disorders such as anxiety and depression (postpartum) occur twice as often in women than in men. The question is what accounts for this big difference between men and women?
2. Most women are emotionally and psychologically traumatized.
3. Women suffer sleep deprivation more than men (starting from puberty, pregnancy, and parturition).
4. Women are caregivers to their children, parents, and even extended family members.
5. Women are over three times more likely than men to be diagnosed with an autoimmune disorder, including those that attack the brain, like multiple sclerosis.
6. Women are up to four times more likely to suffer from migraines and headaches than men.
7. Women are more prone than men to developing meningiomas, the most common brain tumor. Strokes kill more women than men (Lisa Mosconi, The XX Brain).

4.8 Facts about Women and Alzheimer's

The statistics on women and Alzheimer's disease are very startling.

1. Do you know that women are disproportionately affected by Alzheimer's disease more than men?
2. More women are affected by dementia worldwide than men, for every man with dementia there are two women (Mielke et al. 2014).
3. Brain scan tells us that the rate at which brain cells are dying in the brain is faster in women than in men.
4. Women are more likely to live longer than men. However, although the risk increases with age, dementia is caused by the disease of the brain and not age alone.
5. Women in their 60 s are more than twice as likely to develop Alzheimer's disease over the rest of their lives as they are to develop breast cancer (Mielke et al. 2014).
6. Every 66 s someone in the United States develops Alzheimer's, and by 2050, it will be every 33 s; it may interest you to know that two-thirds of them are women.

7. Another disconcerting stat is that once women develop mild cognitive impairment, their cognitive decline is two times faster than men.

4.9 Factors that Place Women at Greater Risk

It's not exactly clear why women are more affected by Alzheimer's than men, but there may be several factors at play. According to Dr. Caldwell, women tend to decline faster than men after receiving a diagnosis for Alzheimer's. I believe that women's vulnerability to Alzheimer's disease is multifactorial and studies should look beyond age and embark on a holistic research to investigate why women are disproportionately affected by Alzheimer's disease. Women typically live longer than men, too, and while the No. 1 risk factor for Alzheimer's is aging, that may not be the whole story (Dr. Caldwell 2022).

4.10 The Factors that Contribute Greatly

Longevity Age is a risk factor. Scientists once thought that women were harder hit by Alzheimer's because of generally living longer than men. Heather Snyder, senior director of medical and scientific operations for the Alzheimer's Association, says that this isn't the case and that new studies suggest there are different biological pathways in women's brains and that hormones or even the way women's brains metabolize food differently may explain why Alzheimer's manifests itself more in women. Again, Alzheimer's is not a natural aging sickness, and studies have shown that Alzheimer's disease is thought to begin 20 years or more before symptoms arise. Women's brains have unique risk factors for dementia that until now have been ignored by science (Mielke et al. 2014).

Female Hormone Estrogen gives a neuroprotective effect to the brain. A decline in estrogen is the leading cause of problems that affect women after age 35. It has been hypothesized that the rapid decrease in estrogen during menopause functions as a trigger for the development of AD in women due to these neuroprotective abilities (Paganini-Hill and Henderson 1994).

Menopause Menopause and estrogen loss are a huge area of investigation for Alzheimer's because estrogen supports an area of the brain (the hippocampus) responsible for forming new memories. It's this part of the brain that's first targeted when Alzheimer's develops, so as women age, they may be even more affected. Plus, women have a greater increase in Alzheimer's risk, compared to men, when they carry a gene associated with late-onset Alzheimer's. But on the other hand, there is a line of research that suggests having two X-chromosomes might put women at an advantage. The menopause state in midlife is associated with a decline in the brain's ability to utilize glucose as its primary fuel and an increase in neuro-inflammatory responses in the brain. That combination is a dual hit for the female brain (Roberta Diaz Brinton n.d.).

APOE e4 Gene It is unclear whether genetic risk operates differently in women and men in the development of, or susceptibility to, Alzheimer's pathology (Carter et al. 2012). Several studies have shown that the APOE-e4 genotype, the best-known common genetic risk factor for Alzheimer's dementia, may have a stronger association with Alzheimer's dementia (Altmann et al. 2014; Ungar et al. 2014) and neurodegeneration (Hohman et al. 2018) in women than in men. A recent meta-analysis found no difference between men and women in the association between APOE-e4 and Alzheimer's dementia overall, although age played an interesting interactive role. That is, APOE-e4 was related to higher Alzheimer's risk in women than men between ages 55 and 70, when APOE-e4 is thought to exert its largest effects (Neu et al. 2017). It is unclear whether the influence of APOE-e4 may depend on the sex hormone estrogen (Yaffe et al. 2000; Kang and Grodstein 2012).

New Gene Linked to the Increased Risk of Alzheimer's in Women Recent research has shown the Genome-wide association studies (GWAS), O^6-methylguanine-DNA methyltransferase, MGMT gene may be associated with a higher risk of AD in two different populations, particularly in women without APOE e4. About 60% of people with AD don't express apolipoprotein E (APOE e4), its most established genetic risk factor. The study found that the expression of MGMT contributes to the development of toxic proteins associated with Alzheimer's, especially in women: To understand if other genes involved in tau-related diseases are related to AD risk in women, the researchers performed GWAS on two different populations:

- Thirty-one members of the Hutterites, a group of people with common ancestry, recognized for their relatively small gene pool, 22 of whom were women.
- 10,340 women without APOE e4, who were part of the Alzheimer's Disease Genetics Consortium (ADGC). These included 3399 AD cases and 6905 controls. They found that in both populations, the MGMT gene was associated with AD risk in women lacking APOE e4 (Chung et al. 2022).

Stress and Sleep Deprivation Another brain region which is affected by the decline in estrogen is the brain stem. The brain stem regulates sleep and stress. Stress increases the cortisol level in the body which leads to a decrease in estrogen, hence sleep deprivation. All these happen because women juggle a lot of things. Puberty, pregnancy, and parturition cause negative changes in a woman's brain. Sleep deprivation as a mother and a caregiver also affects hormones and neuronal connectivity.

Education Women are generally less educated than men. People with more years of formal education are at lower risk for Alzheimer's and other dementias than those with fewer years of formal education (Kukull et al. 2002; Evans et al. 2003, Fitzpatrick et al. 2004, Sando et al. 2008, Stern 2012, Hendrie et al. 2018). The underlying reasons for the relationship between formal education and reduced Alzheimer's risk are unclear. Some researchers believe that having more years of education builds "cognitive reserve." Cognitive reserve refers to the brain's ability to make flexible and efficient use of cognitive networks (networks of neuron-to-neuron connections) to enable a person to continue to carry out cognitive tasks despite brain changes (Stern 2002; Stern et al. 2018). The number of years of formal education is not the only determinant of cognitive reserve. Having a mentally stimulating job and engaging in other mentally stimulating activities may also help build cognitive reserve (Fisher et al. 2014; Grzywacz et al. 2016; Pool et al. 2016; Then et al. 2014).

4.11 Way Forward: Change in Lifestyle Can Reverse or Slow the Early Onset of Alzheimer's

There is increasing evidence that there are steps people can take to reduce their risk of developing certain dementias or to delay their onset. In 2015, the Alzheimer's Association evaluated the state of the evidence on the effects of modifiable risk factors. They concluded that "regular physical activity and management of cardiovascular risk factors (especially diabetes, obesity, smoking and hypertension) reduce the risk of cognitive decline and may reduce the risk of dementia." A healthy diet and lifelong learning/cognitive training may also reduce the risk of cognitive decline. These findings were largely confirmed by the Institute of Medicine in 2015. The possible ways to reduce, reverse, or slow the risk of developing Alzheimer's are as follows:

Early Detection Early detection of Alzheimer's signs and symptoms will be very useful in making better decisions on the treatment plan, which if properly adopted will slow down or reverse the progression of Alzheimer's disease.

Exercise and Stay Physically Active Exercise is a great fuel to the brain because it increases blood flow to the brain and also can build new brain cells. Exercise regularly and be physically active. Physical exercises such as dancing, walking, jogging, biking, swimming, or aerobic classes *at least* 30 min per day, 5 days per week have proven very effective in improving brain function, slowing down the disease progression and reducing brain cell loss in older persons. It reduces the risk of heart-related disease, and thus can also be protective against dementia. It does not have to be strenuous: It is most effective when combined with social activity and a brain-healthy diet. Physical activity that combines mental activity with social activity (e.g., going for a walk and mentally planning the route) may also be effective.

Healthy Diet Eat right and maintain a healthy weight. It is important to cut down on the Western diet and eat more of a Mediterranean-style diet.

- **Reduce intake of foods high in fat or cholesterol**:
  High saturated fat and cholesterol clog the arteries. People living with high cholesterol or high blood pressure were six times as likely to develop AD. "Good" cholesterol, or HDL, may protect the brain (e.g., olive oil). Bake or grill instead of frying foods.
- **Increase your intake of protective foods:**
- Dark-skinned fruits and vegetables have the highest level of antioxidants.
  - Kale, spinach, brussels sprouts, alfalfa sprouts, broccoli, beets, red bell pepper, onion, corn, and eggplant.
  - Prunes, raisins, blueberries, blackberries, strawberries, raspberries, plums, oranges, red grapes, and cherries.
  - Cold water fish contain beneficial omega-3 fatty acids: halibut, mackerel, salmon, trout, and tuna.
  - Nuts can be a useful part of your diet; almonds, pecans, and walnuts are a good source of vitamin E, an antioxidant.
  - Seeds: pumpkin seeds, flaxseeds, chia seeds, sunflower seeds.
  - Seed oil: Flaxseed oil is very good for phyto-estrogen. Eat other foods sparingly.

Taken from: http://www.alz.org/we_can_help_brain_health_maintain_your_brain.asp.

Remain Socially Active Stay socially connected, attend social gatherings occasionally. Recent research suggests that those who combine physical, mental, and social activity are most likely to prevent dementia. Consider doing the following: stay active in the workplace; volunteer in community groups and causes; join bridge clubs, square dancing clubs, or other social groups; and travel.

- **Sleep**: Quality sleep helps to wipe out some Alzheimer's disease plagues.
- Meditation.
- Avoid drinking too much alcohol.
- Stop smoking.

Get Educated to Stay Mentally Active The idea here is to build a better brain, or what Dr. Caldwell refers to as "cognitive reserve." By getting your formal education and continuing to learn and practice what you've learned over time, you're strengthening your brain along the way, almost like a suit of armor. "The better brain you build, the more you have to draw from if you were to ever get a neurodegenerative disorder like Alzheimer's disease," says Dr. Caldwell.

It's possible that pushing yourself to actively learn whenever possible can strengthen your cognitive abilities, so even if you have Alzheimer's, your symptoms may show up later in life, or in the best-case scenario, not at all. Keep your brain active every day, start with something small, then add another change each day. Stay curious and involved—commit to lifelong learning, read, write, work crossword or other puzzles, and attend lectures and plays. Enroll in courses at your local adult education center, community college, or other community groups (Taken from:

http://www.alz.org/we_can_help_brain_health_maintain_your_brain.asp).

4.12 Implication for SDG 3

4.12.1 SDG3, Target 3.3

By 2030, end the epidemics of AIDS, tuberculosis, malaria, and neglected tropical diseases and combat hepatitis, water-borne diseases, and other communicable diseases: Dementia, which was previously seen as a Western world problem, has gradually become a serious stigma for people in the low- and middle-income countries.

4.12.2 Target 3.7

By 2030, ensure universal access to sexual and reproductive healthcare services, including family planning, information and education, and the integration of reproductive health into national strategies and programs: Women are generally less educated than men. People with more years of formal education are at lower risk for Alzheimer's and other dementias than those with fewer years of formal education.

4.12.3 Target 3.8

Achieve universal health coverage, including financial risk protection; access to quality essential healthcare services; and access to safe, effective, quality, and affordable essential medicines and vaccines for all: Early detection of Alzheimer's signs and symptoms will be very useful in making better decisions on the treatment plan, which if properly adopted will slow down or reverse the progression of Alzheimer's disease.

4.12.4 Target 3.9

By 2030, substantially reduce the number of deaths and illnesses from hazardous chemicals and air, water, and soil pollution and contamination: AD is one of the leading causes of death in the world, currently ranked sixth.

4.13 Conclusion

Alzheimer's disease is tagged as an old-age problem and therefore receives little or no attention. Females suffering from this disease are stigmatized and the families are advised not to waste their resources in treatment. The gender disparity in dementia and AD is alarming, hence strategic plans to help women suffering from this condition are highly recommended because dementia is not just a natural aging problem but a progressive distortion in the brain tissues caused by the accumulation of hormonal imbalances, emotional trauma, abuse, and other stress-related conditions that females encounter at different stages of their lifetime.

People of all ages are to live healthy lives and to advance well-being, according to SDG 3. At every stage of one's life, beginning at birth, health and well-being are crucial. However, it is impossible to discuss health and happiness without mentioning mental health. The brain is the central processing unit that steers all bodily functions. Alzheimer's disease is a serious public health problem that challenges the well-being of an individual. A type of dementia that affects older people more frequently is Alzheimer's disease. Memory loss, confusion, personality changes, and other symptoms of dementia can all have an impact on a person's overall health, including the caregiver.

References

Altmann A, Tian L, Henderson VW, Greicius MD. Alzheimer's Disease Neuroimaging Initiative Investigators. Sex modifies the APOE-related risk of developing Alzheimer disease. Ann Neurol. 2014;75(4):563–73.

Bateman RJ, Xiong C, Benzinger TL, Fagan AM, Goate A, Fox NC, et al. Clinical and biomarker changes in dominantly inherited Alzheimer's disease. N Engl J Med. 2012;367(9):795–804.

Beam CR, Kaneshiro C, Jang JY, Reynolds CA, Pedersen NL, Gatz M. Differences between women and men in incidence rates of dementia and Alzheimer's disease. J Alzheimers Dis. 2018;64(4):1077–83. https://doi.org/10.3233/JAD-180141. PMID: 30010124; PMCID: PMC6226313

Braak H, Thal DR, Ghebremedhin E, Del Tredici K. Stages of the pathologic process in Alzheimer disease: age categories from 1 to 100 years. J Neuropathol Exp Neurol. 2011;70(11):960–9.

Roberta Diaz Brinton, PhD, director of the Center for Innovation in Brain Science at the University of Arizona and a leading neuroscientist in the field of Alzheimer's and the aging female brain.

Burns A, Lewis G, Jacoby R, Levy R. Factors affecting survival in Alzheimer's disease. Psychol Med. 1991;21(2):363–70.

Caldwell J. Sex and gender in Alzheimer's disease risk and resilience 2022.

Carter CL, Resnick EM, Mallampalli M, Kalbarczyk A. Sex and gender differences in Alzheimer's disease: recommendations for future research. J Women's Health. 2012;21(10):1018–23.

Chene G, Beiser A, Au R, Preis SR, Wolf PA, Dufouil C, et al. Gender and incidence of dementia in the Framingham Heart Study from mid-adult life. Alzheimers Dement. 2015;11(3):310–20.

Chung J, Das A, Sun X, et al. Genome-wide association and multi omics studies identify MGMT as a novel risk gene for Alzheimer's disease among women. Alzheimers Dement. 2022:1–13.

Daniel WF, Bennett DA, Dong H. Sexual dimorphism in predisposition to Alzheimer's disease. Neurobiol Aging. 2018;70:308–24.

Evans DA, Bennett DA, Wilson RS, Bienias JL, Morris MC, Scherr PA, et al. Incidence of Alzheimer disease in a biracial urban community: relation to apolipoprotein E allele status. Arch Neurol. 2003;60(2):185–9.

Fisher GG, Stachowski A, Infurna FJ, Faul JD, Grosch J, Tetrick LE. Mental work demands, retirement, and longitudinal trajectories of cognitive functioning. J Occup Health Psychol. 2014;19(2):231–42.

Fitzpatrick AL, Kuller LH, Ives DG, Lopez OL, Jagust W, Breitner JC, et al. Incidence and prevalence of dementia in the cardiovascular health study. J Am Geriatr Soc. 2004;52(2):195–204.

Gordon BA, Blazey TM, Su Y, Hari-Raj A, Dincer A, Flores S, et al. Spatial patterns of neuroimaging biomarker change in individuals from families with autosomal dominant Alzheimer's disease: a longitudinal study. Lancet Neurol. 2018;17(3):241–50.

Grzywacz JG, Segel-Karpas D, Lachman ME. Workplace exposures and cognitive function during adulthood: Evidence from National Survey of Midlife Development and the O*NET. J Occup Environ Med. 2016;58(6):535–41.

Hebert LE, Scherr PA, McCann JJ, Beckett LA, Evans DA. Is the risk of developing Alzheimer's disease greater for women than for men? Am J Epidemiol. 2001;153(2):132–6.

Hebert LE, Weuve J, Scherr PA, Evans DA. Alzheimer disease in the United States (2010-2050) estimated using the 2010 census. Neurology. 2013;80(19):1778–83.

Hendrie HC, Smith-Gamble V, Lane KA, Purnell C, Clark DO, Gao S. The Association of early life factors and declining incidence rates of dementia in an elderly population of African Americans. J Gerontol B Psychol Sci Soc Sci. 2018;16(73, suppl 1):S82-9.

Hohman TJ, Dumitrescu L, Barnes LL, Thambisetty M, Beecham G, Kunkle B, et al. Sex-specific association of apolipoprotein E with cerebrospinal fluid levels of tau. JAMA Neurol. 2018;75(8):989–98.

Ibeachu PC, Uahomo PO, David LK. Sexual dimorphism in dementia and Alzheimer diseases—A neuropsychiatric hospital based study. Int Neuropsychiatr Dis J. 2022;17(1):36–45.

Jack CR, Lowe VJ, Weigand SD, Wiste HJ, Senjem ML, Knopman DS, et al. Serial PiB and MRI in normal, mild cognitive impairment and Alzheimer's disease: implications for sequence of pathological events in Alzheimer's disease. Brain. 2009;132:1355–65.

Kang JH, Grodstein F. Postmenopausal hormone therapy, timing of initiation, APOE and cognitive decline. Neurobiol Aging. 2012;33(7):1129–37.

Kua EH, Ho E, Tan HH, Tsoi C, Thng C, Mahendran R. The natural history of dementia. Psychogeriatrics. 2014;14(3):196–201.

Kukull WA, Higdon R, Bowen JD, McCormick WC, Teri L, Schellenberg GD, et al. Dementia and Alzheimer disease incidence: a prospective cohort study. Arch Neurol. 2002;59(11):1737–46.

Lin FC, Chuang YS, Hsieh HM, et al. Early statin use and the progression of Alzheimer disease: a total population-based case-control study. Medicine (Baltimore). 2015;94(47):e2143.

Liu CC, Li CY, Sun Y, Hu SC. Gender and age differences and the trend in the incidence and prevalence of dementia and Alzheimer's disease in Taiwan: a 7-year National Population-Based Study. Biomed Res Int. 2019;5378540 https://doi.org/10.1155/2019/5378540. PMID: 31815145; PMCID: PMC6878786

Mielke MM, Vemuri P, Rocca WA. Clinical epidemiology of Alzheimer's disease: assessing sex and gender differences. Clin Epidemiol. 2014;6:37–48.

Neu SC, Pa J, Kukull W, Beekly D, Kuzma A, Gangadharan P, et al. Apolipoprotein E genotype and sex risk factors for Alzheimer disease: a meta-analysis. JAMA Neurol. 2017;74(10):1178–89.

Niu H, Álvarez-Álvarez I, Guillén-Grima F, Aguinaga-Ontoso I. Prevalence and incidence of Alzheimer's disease in Europe: a meta-analysis. Neurologia. 2017;32(8):523–32. https://doi.org/10.1016/j.nrl.2016.02.016. English, Spanish Epub 2016 Apr 26. PMID: 27130306

Paganini-Hill A, Henderson V. Estrogen deficiency and risk of Alzheimer's disease in women. Am J Epidemiol. 1994;140:256–61.

Podcasy JL, Epperson CN. Considering sex and gender in Alzheimer disease and other dementias. Dialogues Clin Neurosci. 2016;18(4):437–46. https://doi.org/10.31887/DCNS.2016.18.4/cepperson. PMID: 28179815; PMCID: PMC5286729

Pool LR, Weuve J, Wilson RS, Bültmann U, Evans DA, Mendes de Leon CF. Occupational cognitive requirements and late-life cognitive aging. Neurology. 2016;86(15):1386–92.

Rajan KB, Weuve J, Barnes LL, McAninch EA, Wilson RS, Evans DA. Population estimate of people with clinical AD and mild cognitive impairment in the United States (2020-2060). Alzheimers Dement. 2021; https://doi.org/10.1002/alz.12362. Online ahead of print.

Reiman EM, Quiroz YT, Fleisher AS, Chen K, Velez-Pardos C, Jimenez-Del-Rio M, et al. Brain imaging and fluid biomarker analysis in young adults at genetic risk for autosomal dominant Alzheimer's disease in the presenilin 1 E280A kindred: a case-control study. Lancet Neurol. 2012;11(2):1048–56.

Sando SB, Melquist S, Cannon A, Hutton M, Sletvold O, Saltvedt I, et al. Risk-reducing effect of education in Alzheimer's disease. Int J Geriatr Psychiatry. 2008;23(11):1156–62.

Seshadri S, Wolf PA, Beiser A, Au R, McNulty K, White R, et al. Lifetime risk of dementia and Alzheimer's disease. The impact of mortality on risk estimates in the Framingham study. Neurology. 1997;49(6):1498–504.

Sinforiani E, Citterio A, Zucchella C, et al. Impact of gender differences on the outcome of Alzheimer's disease. Dement Geriatr Cogn Disord. 2010;30:147–54.

Stern Y. What is cognitive reserve? Theory and research application of the reserve concept. J Int Neuropsychol Soc. 2002;8:448–60.

Stern Y. Cognitive reserve in ageing and Alzheimer's disease. Lancet Neurol. 2012;11(11):1006–12.

Stern Y, Arenaza-Urquijo EM, Bartres-Faz D, Belleville S, Cantilon M, Chetelat G, et al. Whitepaper: defining and investigating cognitive reserve, brain reserve, and brain maintenance. Alzheimers Dement. 2018; pii:S1552-5260(18): 33491-5

Then FS, Luck T, Luppa M, Arelin K, Schroeter ML, Engel C, et al. Association between mental demands at work and cognitive functioning in the general population: results of the health study of the Leipzig research Center for Civilization Diseases. J Occup Med Toxicol. 2014;9:23.

U.S. Census Bureau. National population projections: downloadable files 2014. Available at: https://www.census.gov/data/datasets/2014/demo/popproj/2014-popproj.html. Accessed December 18, 2021.

Ungar L, Altmann A, Greicius MD. Apolipoprotein E, gender, and Alzheimer's disease: an overlooked, but potent and promising interaction. Brain Imaging Behav. 2014;8(2):262–73.

Villemagne VL, Burnham S, Bourgeat P, Brown B, Ellis KA, Salvado O, et al. Amyloid ß deposition, neurodegeneration, and cognitive decline in sporadic Alzheimer's disease: a prospective cohort study. Lancet Neurol. 2013;12(4):357–67.

Yaffe K, Haan M, Byers A, Tangen C, Kuller L. Estrogen use, APOE, and cognitive decline: evidence of gene-environment interaction. Neurology. 2000;54(10):1949–54.

World Health Organization. Geneva, Switzerland: World Health Organization; 2021. Fact sheets of dementia [Internet] [cited 2022 Apr 13]. Available from: https://www.who.int/news-room/factsheets/detail/dementia. [Google Scholar].

Leveraging Artificial Intelligence Technology for Effective Early Diagnosis: Heart Issues

5

Laeticia Nneka Onyejegbu

5.1 Introduction

Artificial intelligence (AI) is the science of making machines do things that would require intelligence if done by men. AI can also be defined as the ability to make computers or machines learn to solve problems that would otherwise require human effort. The first concept of AI was proposed in 1956, by John McCarthy who defined AI as a branch of computer science concerned with making computers behave like humans. The main aim of AI in technology is to make computer systems and machines imitate human behaviour when executing tasks. Advances in computing power have made it possible to analyse large amounts of data quickly with consistency and accuracy using AI. As such, AI has expanded to almost every facet of modern life, including healthcare.

Artificial intelligence has been incorporated into the field of cardiovascular medicine and is increasingly employed to revolutionize the diagnosis, treatment, risk prediction, and clinical care. Heart failure (HF) has a high prevalence, and the mortality rate following hospitalization is quite high. Early detection of HF is of vital importance in shaping the medical and surgical interventions specific to HF patients. This has been accomplished with the advent of the neural network (NN) model, the accuracy of which has proven to be 85% (Farah et al. 2021). AI is used in analysing raw image data from cardiac imaging techniques (such as echocardiography, computed tomography, and cardiac MRI amongst others) and electrocardiogram recordings through the incorporation of an algorithm. Machine learning algorithms such as decision trees, neural network, and logistic regression methods used to build a decision-making model to diagnose congestive HF, and the role of AI in the early detection of future mortality has played a vital role in optimizing cardiovascular disease outcomes.

5.2 Machine Learning

Machine learning (ML) is a subset of AI made up of special algorithms that can automatically improve themselves impressively over time through experience. The dynamic learning nature of a machine learning system has made it appropriate for various applications, and a new way, to solve problems in a constantly changing environment which allows a machine to automatically learn from past data without programming explicitly. The goal of ML is to allow machines to learn from data so that they can give accurate output. Mitchell (1997) defined machine learning as a computer program that learns from experience (E)

L. N. Onyejegbu (✉)
Computer Science Department, Faculty of Computing, University of Port Harcourt, Port Harcourt, Rivers State, Nigeria
e-mail: laeticia.onyejegbu@uniport.edu.ng

E. O. Nwaichi (ed.), *Modernity in Health and Disease Diagnosis: The Account from STEM Women*, Sustainable Development Goals Series, https://doi.org/10.1007/978-3-031-34963-8_5

with respect to some class of tasks (T) and performance measure (P), if its performance at tasks in T, as measured by P, improves with experience E. This definition can be related to the growth of a child in terms of behaviour. As a child grows, there is an increase in its experience E in accomplishing task T, which results in a higher performance measure (P). Arthur Samuel in 1959, defined machine learning as defined machine learning as the subfield of computer science that gives computers the ability to learn without being explicitly programmed. In this definition, computers are not programmed with specific rules, but they learn from huge amounts of datasets fed into them in the domain of interest which allows them to learn the patterns therein and gain experience to solve problems in the same domain. The acquisition of experience starts from learning.

As shown in Fig. 5.1, machine learning technique is categorized into supervised, unsupervised, and reinforcement learning. Supervised learning is used to solve classification and regression problems using categorical and continuous data and is task-driven. Unsupervised learning is used to solve clustering problems by dividing individuals into groups with similar characteristics and is data-driven. While reinforcement learning is used for decision-making and it learns from mistakes.

5.2.1 Supervised Learning

Supervised learning is the search for algorithms that reason from externally supplied instances to produce general hypotheses, which then make predictions about future instances. One of the tasks most frequently carried out by intelligent systems is supervised learning (Osisanwo et al. 2017). It is called supervised learning because the process of an algorithm learning from the training dataset can be thought of as a teacher supervising the learning process. In supervised learning, the algorithm "learns" from the training dataset by iteratively making predictions on the data and adjusting for the correct answer. To "learn" is the process through which an algorithm modifies itself to being able to produce a certain result with a given input. Supervised learning involves the training of the model on a labelled dataset. There are two types of supervised learning: classification and regression.

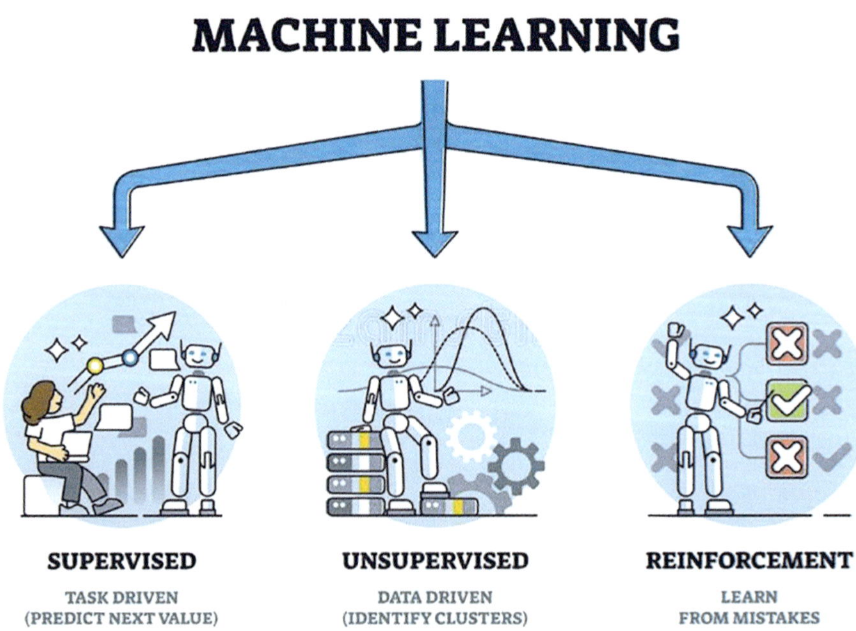

Fig. 5.1 Machine learning techniques (Source: https://www.dreamstime.com/about-stock-image-licenses)

5.2.2 Unsupervised Learning

Unsupervised learning involves the training of a model in an unlabelled dataset. The model learns on its own by learning the features of the training dataset. Based on those learning features, the model makes predictions on test data. Unsupervised learning aims to reveal hidden patterns in data. There are several types of unsupervised learning approaches and algorithms, and they include clustering, k-means to agglomerative, principal component analysis, and fuzzy C-means (Salim D, 2021).

5.2.3 Reinforcement Learning

Reinforcement learning is the type of learning guided by a specific objective. An agent learns by interacting with an unknown environment, typically in a try-and-error way. The agent receives feedback in terms of a reward (or punishment) from the environment; then, it uses this feedback to train itself and collect experience and knowledge about the environment (Muddasar et al. 2020).

5.3 Deep Learning

Deep learning is a subfield of machine learning inspired by the way biological neurons in the human brain work. Hence, the concept of artificial neural networks (ANNs) is the core foundation on which deep learning models are based. Figure 5.2 shows the performance comparison between deep learning and other learning algorithms. Deep learning performs better than traditional machine learning algorithm when trained with large amounts of data. However, model training time and model interpretability are still some of the factors slowing down the adoption of deep learning over the traditional machine learning algorithm in some of today's business environments.

Deep learning algorithms take longer time to train than traditional machine learning algorithms. The main reason for this is because there are many parameters used in deep learning and often, large datasets (in thousands and millions) are used for training the model and the training process involves many iterations in order to optimize performance. Another factor slowing down the adoption of deep learning is interoperability. Interoperability focuses on getting an explanation of how the model makes decisions. This is vital especially in a business environment because it aids trust in the model's decision by the various stakeholders. Deep learning has been applied to diverse fields such as speech recognition, social network filtering, bioinformatics, drug design, and medical image interpretation. Deep neural network is an example of a deep learning algorithm. It comprises of series of layers: an input layer, a cascade of processing units or hidden layers, and an output layer. Each of the layers comprises of individual neurons that extract and transfer data in a hierarchical fashion into more composite representations. Data from one layer is processed and fed into the next layer. Different types of neural networks have been developed; the type of neural network employed depends on the type and complexity of the analysis being performed (Kraus et al. 2020).

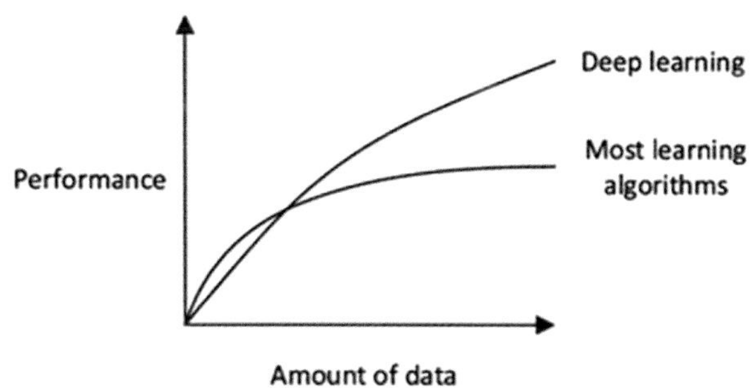

Fig. 5.2 Performance of deep learning vs. other machine learning algorithms based on the amount of data (Kraus et al. 2020)

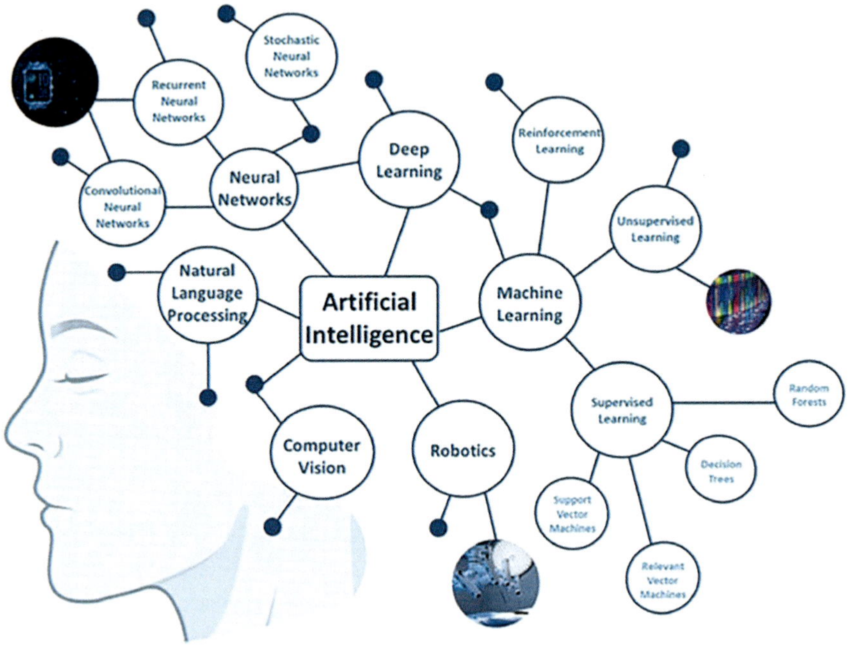

Fig. 5.3 Application fields of artificial intelligence (Konstantina et al. 2022). https://www.mdpi.com/2673-7426/2/4/49

5.4 Technical and Application Fields of Artificial Intelligence

Figure 5.3 shows some of the algorithms used in solving machine learning and deep learning problems such as the neural network, random forest, decision trees, support vector machines, recurrent neural networks, convolutional neural networks, and stochastic neural networks. The areas of AI technology are robotics, natural language processing, computer vision, and others. There are five schools of thought that AI development was based on, and they are Bayesian, connectionism, symbolism, evolutionism, and analogizers. Artificial intelligence can also be used for smart finance, smart education, smart healthcare, smart city, smart agriculture, smart home, and so on.

5.5 Overview of the Heart

The heart has been considered as the king of the body which supplies blood to different organs. The heart is made up of four chambers: the left and right atrium and the left and right ventricles. It is protected and anchored to the chest by a double-walled sac called the pericardium. The different chambers of the heart are connected by valves. Valves also connect your heart to the rest of the body. Blood is circulated by the heart through two pathways: the pulmonary circuit and the systemic circuit. The pulmonary circuit runs deoxygenated blood through the right ventricle to the lungs, where it becomes oxygenated before returning to the left atrium. In the systemic circuit, the oxygen-infused blood leaves through the left ventricle into the aorta. From there, it enters the arteries and capillaries which supply one's tissue with oxygen. The deoxygenated blood then comes back to the heart and the process starts anew. As a muscle, the heart uses some of the oxygen as well, which means it has its own arteries. When these arteries experience blockage, a heart attack can happen. It can also cause damage and scarring to the heart. Improper performance can also cause a person to go into cardiac arrest.

5.6 Artificial Intelligence in Cardiology

Cardiology is a branch of medicine that specializes in diagnosing and treating diseases of the heart, blood vessels, and circulatory system. Heart disease refers to any condition affecting the heart. There are many types of heart diseases, some of which are preventable. AI can be used to program computers to process and respond to data quickly and consistently for better treatment outcomes, and this includes detecting heart diseases. AI can save time and improve reproducibility because the machines will do the work the same way, every time.

5.7 Application of Artificial Intelligence in Heart Attack Diagnosis

Kagiyama et al. (2019) developed a machine learning model to predict patients' long-term risk of a heart attack. This was done by combining coronary artery calcium scoring with non-contrast computed tomography, which indicates the accumulation of cholesterol within artery walls. Cardiologists can tell from the timing of the heartbeat in scans if there is a problem. This has saved a lot of patients from either being sent home and having a heart attack or undergoing an unnecessary operation.

Scientists at Google discovered a new way to assess a person's risk of heart disease using machine learning. This was done by analysing scans of the back of a patient's eye. The company's software can accurately deduce data, including an individual's age, blood pressure, and whether they smoke or not. This can then be used to predict their risk of suffering a major cardiac event such as a heart attack with roughly the same accuracy as current leading methods. The algorithm potentially makes it quicker and easier for doctors to analyse a patient's cardiovascular risk, as it does not require a blood test (Vincent 2018).

5.8 Application of Artificial Intelligence in Hypertrophic Cardiomyopathy and Cardiac Amyloidosis Diagnosis

Hypertrophic cardiomyopathy is a disease that causes the heart muscle to thicken and stiffen. As such it prevents the heart from relaxing and blocks blood flow out of the heart, resulting in damage to heart valves and fluid build-up in the lungs, making it harder for the heart to work and pump out blood, and causing abnormal heart rhythms. Figure 5.4 shows a normal heart tissue and the heart tissue with thickened ventricular septum, resulting in hypertrophic cardiomyopathy.

Cardiac amyloidosis is a disorder caused by deposits of an abnormal protein (amyloid) in the heart tissue as shown in Fig. 5.5. As amyloid builds up, it takes the place of the healthy heart muscle, making it difficult for the heart to work properly. Cardiac amyloidosis often goes undetected because patients might not have any symptoms, or they might experience symptoms only sporadically. The disease tends to affect older, black men or patients with cancer or diseases that cause inflammation.

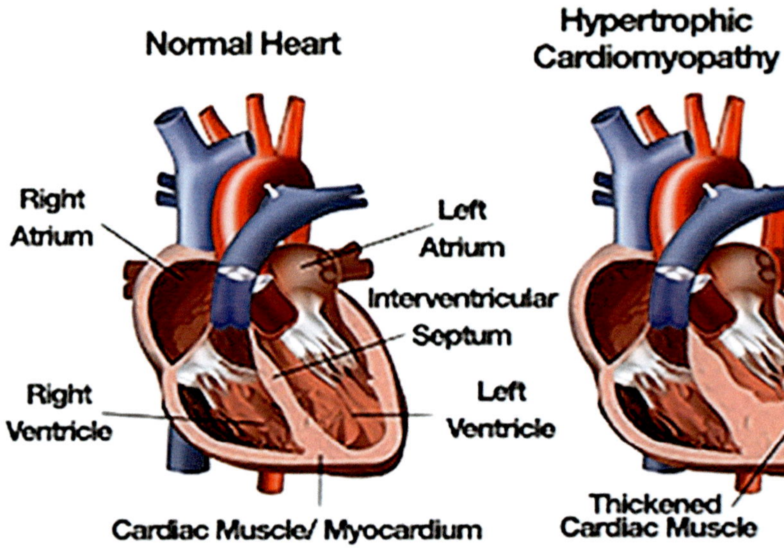

Fig. 5.4 Thickened ventricular septum (Source: https://utswmed.org/medblog/hypertrophic-cardiomyopathy/)

Fig. 5.5 Deposits of amyloid between heart cells (Source: https://www.google.com/imgres)

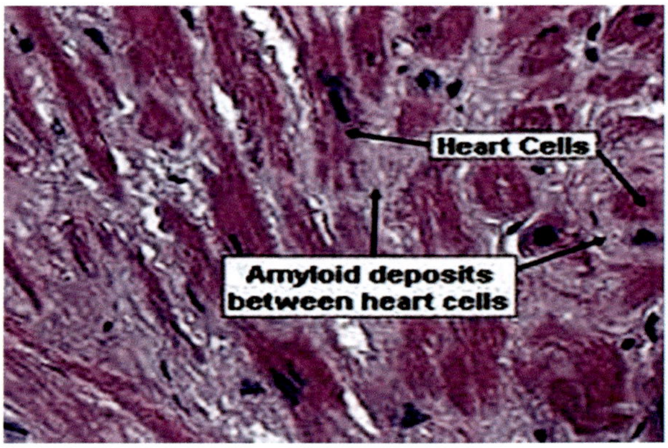

Duffy et al. (2022) developed a new AI algorithm that can effectively identify and distinguish between the two life-threatening heart conditions that are often difficult to diagnose: hypertrophic cardiomyopathy and cardiac amyloidosis. These two heart conditions are challenging for even expert cardiologists to accurately identify, and so patients often go on for years before receiving a correct diagnosis. According to Duffy et al. (2022), the AI algorithm they developed can pinpoint disease patterns that cannot be seen by the naked eye, and these patterns are used to predict the right diagnosis. The two-step, novel algorithm was trained on over 34,000 cardiac ultrasound videos from Cedars-Sinai and Stanford Healthcare's echocardiography laboratories. When applied to these clinical images, the algorithm identified specific features related to the thickness of heart walls and the size of heart chambers to efficiently flag certain patients as suspicious for having potentially unrecognized cardiac diseases. The AI algorithm identified high-risk patients with more accuracy than the well-trained eye of a clinical expert. This is because the algorithm picks up subtle cues on ultrasound videos that distinguish between heart conditions that can often look very similar to more benign conditions, as well as to each other, on initial review. Without comprehensive testing, cardiologists find it challenging to distinguish between similar appearing diseases and changes in heart shape and size that can sometimes be thought of as a part of normal aging. One of the most important aspects of this AI technology is not only the ability to distinguish abnormal from normal conditions, but also to distinguish which underlying potentially life-threatening cardiac conditions may be present with warning signals that are now detectable, well before the disease clinically progresses to the point where it can impact health outcomes, because the treatment and management of each cardiac disease are very different, early diagnosis enables patients to begin effective treatments, prevent adverse clinical events, and improve their quality of life—a deliverable of SDG 3. Cardiac amyloidosis and hypertrophic cardiomyopathy often look very similar to each other on an echocardiogram, the most used cardiac imaging diagnostic (Duffy et al. 2022).

5.9 Conclusion

Artificial intelligence can help doctors in early diagnosis and treatment decisions. Estimating the socioeconomic impact of AI on health systems is fundamental to advancing the current discourse on the role AI can and should have in the health sector. With significant challenges lying ahead, such as an ageing population, growing demand for services, increased costs, and healthcare staff shortages, this industry should embrace AI. The speed of innovation in AI applications for the healthcare industry is increasing. A thriving ecosystem of start-ups, large medical technology players, and "Big Tech" companies are starting to roll out AI-enabled solutions. AI technology will help in early risk prediction and diagnosis of serious and complex heart problems, thereby saving lives.

References

Duffy G, Paul P, Cheng NY, He B, Kwan AC, Shun-Shin MJ, Alexander KM, Ebinger J, Lungren MP, Rader F, Liang DH, Ouyang D, Schnittger I, Ashley EA, Zou JY, Patel J, Witteles R, Cheng S. High-throughput precision phenotyping of left ventricular hypertrophy with cardiovascular deep learning. JAMA Cardiol J. 2022;7(4):386–95.

Farah Y, Syed M, Aisha N, Adina J, Sana K, Sarush AS, Pankaj K, Shiza S, Syed AH, Chandrashekhar D, Ali S, Ahmad M, Sanchit C, Hassan M. Artificial intelligence in the diagnosis and detection of heart failure: the past, present, and future. J Rev Cardiovasc Med. 2021;22(4):1095–113.

Kagiyama N, Sirish S, Farjo PD, Partho PS. Artificial intelligence: practical primer for clinical research in cardiovascular disease. J Am Heart Assoc. 2019;8(17):1–12.

Konstantina A, Glykeria ND, Panagiotis GA, Andreas S. Artificial intelligence: the milestone in modern biomedical research. J BioMed Inform. 2022;2(4):727–44.

Kraus M, Stefan F, Asil O. Deep learning in business analytics and operations research: models, applications, and managerial implications. Eur J Oper Res. 2020;281:628–41.

Mitchell. What is machine learning. 1997 Retrieved 23 May 2022 from https://towardsai.net/p/machine-learning/what-is-machine-learning-ml-b58162f97ec7.

Muddasar N, Syed TH, Antonio C. A gentle introduction to reinforcement learning and its application in different fields. IEEE Open Access J. 2020;8:1–25.

Osisanwo FY, Akinsola JET, Awodele O, Hinmikaiye JO, Olakanmi O, Akinjobi J. Supervised machine learning algorithms: classification and comparison. Int J Comput Trends Technol (IJCTT). 2017;48(3):128–38.

Salim D. Unsupervised learning—a systematic literature review. 2021 Retrieved on April 4, 2022, from https://www.researchgate.net/publication/357380639

Vincent J. *Nature Journal of Biomedical Engineering* 2018, https://www.theverge.com/authors/james-vincent

6 Antimalarial Drug Resistance and Vulnerable Groups

Ifeyinwa N. Chijioke-Nwauche
and Amaka M. Awanye

6.1 Introduction

Malaria is a protozoan infection in humans caused by the parasite Plasmodium which until recently was thought to be classified into four species: *Plasmodium falciparum, Plasmodium ovale, Plasmodium malariae* and *Plasmodium vivax*. However, further study (Sutherland et al. 2010) has shown that there are two non-recombining species of *ovale* (*ovale curtisi* and *ovale wallikeri*) which are non-sympatric in nature (Oguike et al. 2011). Of all the species, *Plasmodium falciparum* is the agent of the most malignant form of malaria, usually presenting with severity mostly in children in sub-Saharan Africa (Urdaneta et al. 2001). It is the most dangerous form of malaria with the highest rates of complications. It is also the commonest species in virtually all parts of Africa accounting for up to 98% of the confirmed cases in Nigeria and is associated with significant morbidity and mortality. *Plasmodium falciparum* is responsible for virtually all the features of severe malaria. *P. malariae* usually occurs as a mixed infection with *P. falciparum* (NPC, NMCP and ICP 2012). The main vector of malaria in Nigeria is *Anopheles gambiae* but *Anopheles funestus* and *Anopheles arabiensis* are also commonly encountered. *Anopheles melas* is found in the coastal areas (NPC, NMCP and ICP 2012).

Malaria remains a major and important health problem facing developing countries especially in Africa where it has been associated with great morbidity and mortality. Even though half of the world's population is at risk of malaria, it is essentially a disease of the tropics and subtropics particularly the sub-Saharan African region. Malaria cases have also been reported in temperate areas as a result of human migration from tropical countries. The African region carries a disproportionately high share of the global malaria burden, with 94% of malaria cases and deaths in 2019.

The current World Malaria Report 2020 from the World Health Organization shows an estimated 229 million cases of malaria globally in 87 endemic countries, a decline from 238 million in 2000. The report also established a 60% decrease in mortality over the period of 2000–2019, from 736,000 in 2000 to 409,000 in 2019. The African region was home to 94% of malaria cases and deaths in 2019. About 95% of global malaria deaths occurred in 31 countries and about half of these deaths occurred in Nigeria (23%), the

I. N. Chijioke-Nwauche (✉)
Department of Clinical Pharmacy and Management, Faculty of Pharmaceutical Sciences, University of Port Harcourt, Port Harcourt, Rivers State, Nigeria
e-mail: ifeyinwa.chijioke-nwauche@uniport.edu.ng

A. M. Awanye
Department of Pharmaceutical Microbiology and Biotechnology, Faculty of Pharmaceutical Sciences, University of Port Harcourt, Port Harcourt, Rivers State, Nigeria
e-mail: amaka.awanye@uniport.edu.ng

E. O. Nwaichi (ed.), *Modernity in Health and Disease Diagnosis: The Account from STEM Women*, Sustainable Development Goals Series, https://doi.org/10.1007/978-3-031-34963-8_6

Democratic Republic of Congo (11%), the United Republic of Tanzania (5%), Mozambique (4%), Niger (4%) and Burkina Faso (4%). At the beginning of 2020, malaria deaths were at the lowest point ever (World Malaria Report 2020).

Earlier reports show an estimated 863,000 malaria deaths, 767,000 (89%) of which occurred in Africa where malaria is the leading cause of mortality in children under 5 years. However, the World Malaria Report 2011 indicates that over a period of two years, malaria cases had reduced from 243 million cases in 2008 to 216 million in 2010 and the number of deaths had reduced to 655,000 (WHO 2011). Despite this reduction, the global burden of malaria has remained very high, especially in the tropics.

The burden of malaria on the endemic countries transcends beyond the health problems, as it has been associated with hampering the development with a high proportion of the wealth of the nation being drained by the disease. About 8.8 million US dollars loss has been attributed to malaria in the form of treatment costs, prevention and loss of person-hours annually (FMOH 2005). Actions taken to reduce malaria transmission as well as antimalarial drug resistance will work towards achieving the SDG3 goals.

6.2 Treatment of Malaria

SDG3 calls for good health and well-being of people. Drug treatment of malaria is the most common and possibly one of the most important measures for the control of malaria. The goals of treatment of uncomplicated malaria are: to provide a rapid and long-lasting cure, to reduce morbidity including malaria-related anaemia, to prevent the progression of uncomplicated malaria to severe and potentially fatal diseases, and to minimize the likelihood and rate of development of drug resistance in addition to reducing transmission (Sutherland et al. 2005; Sawa et al. 2013).

Recommended treatment for malaria according to the World Health Organization is the use of Artemisinin Combination Therapy (ACT), which involves the use of drugs containing an artemisinin-based drug with a partner drug with another mechanism of action so as to ensure the complete killing of the parasite in the bloodstream. Some of the partner drugs include: lumefantrine, amodiaquine, piperaquine, mefloquine, etc.

However, sometimes the desired treatment goal is not always achieved as a result of treatment failure. According to the National Guidelines for diagnosis and treatment of malaria (2015), causes of treatment failure include:

- Incorrect dosing of the drugs.
- Poor adherence to treatment.
- Poor quality of drugs.
- Interactions with other drugs.
- Poor absorption of drugs.
- Misdiagnosis of the patient.
- Drug resistance.

Among the many factors implicated in treatment failure, the present discussion is focused on drug resistance as a major cause of treatment failure.

6.3 Antimalarial Drug Resistance

The World Health Organization defines antimalarial drug resistance as 'The ability of a parasite strain to survive and multiply despite the proper administration and absorption of an antimalarial drug in the doses equal to or higher than usually recommended, provided drug exposure is adequate' (WHO 2015; WHO 2006). It is a shift to the right of the dose-response curve, thus requiring higher drug concentrations to achieve the same parasite clearance (White 2004). In many cases even the higher concentration results in treatment failure.

The implications of antimalarial drug resistance are such that there is continued transmission of drug-resistant parasites thereby limiting the efforts to control malaria. The spread of resistance through the population is facilitated through increased gametocyte carriage that is transmitted upon new infection (Price et al. 1999). Higher rates of transmission of drug-resistant parasites are achieved through selective advantage which parasites carrying resistant genes develop as a result of the continued use of a drug with the prevalence of resistance in a locality (Handunnetti et al. 1996; Sutherland et al. 2002). Results have shown that the global burden of malaria in sub-Saharan Africa is predominantly maintained by antimalarial drug resistance (Barnes and White 2005). The resultant effect is increased difficulty in the management of malaria.

6.4 Causes of Antimalarial Drug Resistance

Widespread and indiscriminate use of antimalarial drugs exerts a strong selective drug pressure on the malaria parasites to develop high levels of resistance. This indiscriminate use includes self-medication, misdiagnosis, use of sub-optimal doses of the drugs, fake and substandard drugs, easily degradable drugs leading to poor bioavailability of the drugs, and use of monotherapy of the artemisinin drugs (Chijioke-Nwauche et al. 2021). Other factors are host factors (age, immune status, co-morbidity especially with HIV); spontaneous mutations in the parasite gene; parasite mutation rate; overall parasite load; mobile populations and migrants with resultant imported malaria; high treatment costs; and poor adherence to malaria treatment (Djimde et al. 2003).

Influencing factors to antimalarial drug resistance include parasite nature, pharmacological properties of the drug, host genetic factors particularly the immune status of the person and also age (Travassos and Laufer 2009). Individuals with lower immunity like HIV-positive patients, children and pregnant women are more vulnerable to antimalarial resistance. Reduced immunity allows the survival of a residuum of parasites thereby potentially increasing the development, intensification and spread of resistance (Byakika-Kibwika et al. 2010). Additionally, delayed cure rate and higher rate of recrudescence which occur in HIV-positive individuals accelerate the spread of resistant parasites and increase the parasite biomass in both symptomatic and asymptomatic carriers (Birku et al. 2002; Shah et al. 2006; Van Geertruyden et al. 2006). Host genetic factors have been shown to underlie some differences in resistance to malaria, and this has been observed between ethnic groups who live in the same area. It has been observed that the Fulanis have a lower incidence of classic malaria resistance genes than other sympatric ethnic populations (Quinn et al. 2017).

6.5 Vulnerable Groups

Vulnerable groups to malaria include pregnant women (especially in first pregnancies and it is worse in the second and third trimesters), children especially those under 5 years of age, immunocompromised persons [HIV-positive patients (a greater percentage are women)], transplant patients, patients that are undergoing cancer treatment, sickle cell anaemia patients, visitors from the non-endemic region and persons that have undergone splenectomy. The spleen controls the removal of infected red blood cells, and it is the first organ that generates immune response to malaria and usually gets enlarged during infection (del Portillo et al. 2012).

Malaria has been shown to have a long-term effect on cognitive function and educational attainment in children (Jukes et al. 2006; Clarke et al. 2008). Malaria is the commonest cause of fever and death especially in young children (WHO 2000). This is made worse by the situation of multi-drug resistance as a result of self-medication, sub-optimal doses of antimalarial drugs, improper diagnosis, and improper treatment due to resource limitations as is the case in Nigeria and many other sub-Saharan African countries.

In as much as malaria is not gender specific apart from the peculiar vulnerability of pregnant women, there are influencing factors to vulnerability in women, and these include social, economic, and cultural factors and access to preventive and treatment measures. There is marginalization of women due to entrenched inequalities in areas such as education; women are less informed and have less access to treatment and preventive measures due to their low economic power and dependence on their husbands. This is particularly so in rural areas where there are more economically disadvantaged and low social status women (UNDP 2015). All these, in addition to the social pressure of providing meals for the family even when they are ill-disposed and their care-giving responsibility, predispose women more to malaria infection.

Malaria is the leading cause of morbidity and mortality in pregnant women in endemic regions such as Africa and Asia. The consequences of malaria in pregnancy include anaemia which causes about 10,000 maternal deaths, low birth weight babies due to intrauterine growth retardation, abortion, premature delivery, stillbirth and maternal death. About 11% of maternal deaths in Nigeria have been attributed to malaria. As a result of this, the WHO recommends a treatment measure to prevent malaria in pregnancy termed "intermittent preventive treatment for malaria in pregnancy" (IPTp). This is the administration of the drug Sulphadoxine-pyrimethamine (SP) to all pregnant women as part of their antenatal care usually starting in the second trimester. Each dose is given 1 month apart with the aim of taking at least three doses before delivery (WHO 2006). It is recommended that malaria be treated after a definitive or confirmed diagnosis based on tests either using a rapid diagnostic test kit (RDT) or microscopic confirmation of parasites from blood film.

Malaria is associated with many problems in pregnancy, and this includes maternal death, anaemia, abortion, poor fetal growth, low birth weight with consequences of child growth retardation and poor cognitive outcomes, stillbirth, and premature delivery which is more severe in primiparous (first time delivery) especially in partially or non-immune persons (World Malaria Report 2020).

6.6 Resistance and Immune Status of the Patient

The immune system plays a very important role in defence against the infections that attack the body, and malarial infection is not exempt from this. The human immune system is comprised of innate or natural and adaptive immunity. Adaptive/acquired immunity is antigen-specific, which develops after exposure to infection and provides long-lasting protection depending on the host, the type and the number of infections. The responses of adaptive immunity are essentially carried out by white blood cells (WBCs) which are made up of the lymphocytes, monocytes, granulocytes, basophils, neutrophils and eosinophils. In malaria infection, the T lymphocytes are the most important and are activated when a person is exposed to the antigens. The predominant T lymphocytes are the CD4 cells and are the major cells that control blood-stage malaria infection (Perlmann and Troye-Blomberg 2000). The CD4 cells are therefore seen as the bedrock of the body's immune system because upon exposure to malaria infection, they are primed for protection by initiating the recruitment and release of the other cells and chemicals that prevent the pathology of the disease. Furthermore, in HIV patients the CD4 cells are depleted leading to suppressed immunity, and this is worsened in malaria infection. HIV patients were more likely to have PCR-detectable parasitemia when compared to HIV-negative persons (Chijioke-Nwauche et al. 2013). Immunity reduces the chances of survival of resistant parasites by serving as a strong restraint on the emergence of resistance (WHO 2015).

The severity of the malaria disease is dependent on the speed of the response of the immune system. Individuals without pre-existing malaria immunity are at risk of malaria infection whereas in malaria-endemic areas where there is moderate or intense transmission, immunity that is acquired by virtue of several exposures plays a very important role in protecting the individuals from developing the clinical and severe form of the disease (Bates et al. 2004; Cohen et al. 2005). Therefore, over time individuals infected with malaria acquire partial protection and therefore have a reduced risk of severe malaria.

Resistance to antimalarial drugs by *Plasmodium falciparum* parasite has become a major health problem since the first resistance to Chloroquine was recorded in Thailand and Cambodia in the late 1950s. This has spread to other malaria-endemic regions and against other antimalarial drugs like sulphadoxine-pyrimethamine (SP) and mefloquine. Studies (Noedl et al. 2008) reporting artesunate-resistant malaria in western Cambodia and reduced in vivo susceptibility of artesunate (Dondorp et al. 2009) and in vitro resistance to its derivative artemether (Jambou et al. 2005) indicate a great threat to the management of malaria since artemisinin and its derivatives are currently the mainstay for the treatment of uncomplicated malaria. More current reports of resistance to the latest generation partner drug piperaquine used in combination with dihydroartemisinin in Cambodia have led to increasing rates of treatment failure (Duru et al. 2016).

6.7 Mechanisms of Antimalarial Drug Resistance

Antimalarial drug resistance can result either from changes in drug accumulation or efflux or reduced affinity of the drug target resulting from point mutations in the respective genes encoding the target (White et al. 1999). Genetic polymorphisms in one or more genes that do not actually encode the drug target itself but affect drug efflux can lead to reduced drug concentrations within the parasites as in chloroquine, amodiaquine, quinine, mefloquine and halofantrine thereby resulting in drug resistance (Valderramos and Fidock 2006). The selection of parasites with genetic mutations gives rise to antimalarial drug resistance. Polymorphism involves nucleotide change of some of the base pairs in a DNA sequence possibly leading to amino acid change. When there is a change in a single base pair compared to the "common" or "wild type" sequence it is called single nucleotide polymorphism (SNP). These polymorphisms could be synonymous where the nucleotide change does not result in amino acid change, whereas a non-synonymous polymorphism involves an amino acid change. This mutation can cause a significant change in the gene which can result in resistance to the drug by the parasite. Citing an example from an unpublished study on resistance to antimalarial drugs, a nucleotide change from Cytosine to Thymine at position 290 of the gene resulted in amino acid change from Serine to Lysine. The consensus frame *Plasmodium falciparum* multi-drug resistance (*pfmdr*) 3D7 is the *Plasmodium falciparum* wild type gene showing nucleotide Cytosine (C) at position 290 while on Sample 059 frame it has changed to Thymine (T). This change from Cytosine to thymine resulted in amino acid change from Serine (S) to Lysine (L) at position 97, hence it is depicted as S97L (Fig. 6.1). Similarly, in sample 207A the change at nucleotide position 218 from Thymine (T) to Cytosine (C) resulted in the amino acid change of Phenylalanine (F) to Serine (S) depicted as F73S.

Another mechanism of resistance is through altered affinity for the drug target caused by single or multiple point mutations in genes that encode the drug target as in pyrimethamine, cycloguanil, sulphonamide and atovaquone (Wang et al. 1997). Resistance can also be due to the expression of higher levels of the gene through amplification thereby resulting in increased copy number. Antimalarial drug resistance can also occur where there has been no prior resistance in the parent drug because of inadequately treated biomass infections (White and Pongtavornpinyo 2003) or reduced sensitivity to a given drug or class of drugs which occurs as a result of spontaneous mutations (Bloland 2001) and subsequent spread as a result of survival and multiplication. Where a large population of parasites are exposed to drug pressure, resistance is reported to develop more quickly (Farooq and Mahajan 2004).

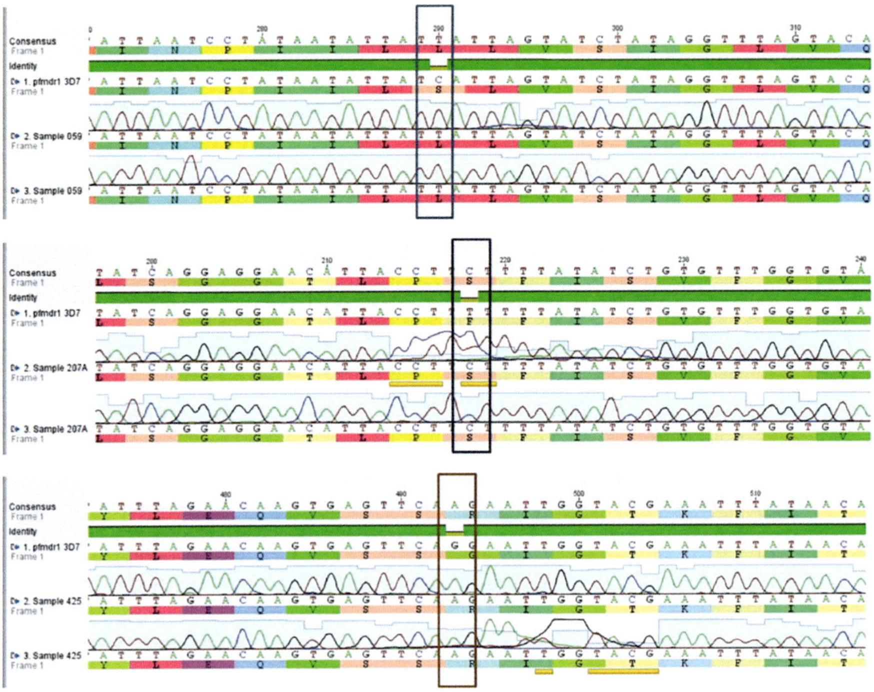

Fig. 6.1 Chromatograms showing examples of non-synonymous mutations from nucleotide changes in a genome of *Plasmodium falciparum* [Source: Unpublished data from PhD thesis of Ifeyinwa Chijioke-Nwauche: Use of Artemether-Lumefantrine in the treatment of asymptomatic malaria in HIV-positive and HIV-negative Nigerian adults in London School of Hygiene and Tropical Medicine, 2014].

6.8 Detecting Resistance

Detection of drug resistance to antimalarial drugs can be achieved through different ways. These include animal model studies, in vitro studies which involve use of drug assays, in vivo studies usually regarded as the gold standard and molecular characterization (Hiasindh and Subhash 2016). Molecular markers of drug-resistant malaria are based on genetic changes that confer parasite resistance to drugs used to treat and prevent malaria (Plowe et al. 2007). They have been proven to be tools for surveillance of resistance and provide additional data that complement clinical observations of the in vivo efficacy of a drug and have been instrumental to policy making with regards to the control of the malaria epidemic (Djimde et al. 2001; Mugittu et al. 2004). They have also served as monitoring tools in parasite drug susceptibility following a change in treatment policy (Laufer et al. 2006).

6.9 Implications of Antimalarial Drug Resistance to Public Health

- Increase in malaria transmission.
- Frequency in severe illness especially in vulnerable groups.
- Increased treatment costs.
- Higher parasite burden which leads to a higher likelihood of resistant parasites.
- Frequency of intake of antimalarials may increase the risk of adverse drug reactions, drug pressure on the parasites and subsequent spread of drug resistance.

Antimalarial drug resistance is a major public health challenge in endemic countries, and the impact is not easily quantifiable. High drug pressure on the parasites which comes as a result of treatment of acute manifestations of the disease rapidly propagates resistance in low transmission areas. However, in areas of high transmission, the manifestation of resistance usually presents clinically, and there is increasing risk of severe anaemia as a result of prolonged or chronic infections. Reports have shown that when resistance develops, there is an associated 2–11-fold increase in malaria mortality in African children (Björkman and Bhattarai 2005). Other clinical consequences of antimalarial drug resistance are increased treatment failure, reduced treatment efficacy, delayed initial therapeutic response, complications such as anaemia and complications during pregnancy.

The World Health Organization advocates that to stop the spread of resistant parasites, in areas with no known resistance, control measures should be used to reduce transmission. Also, preventive measures to reduce transmission such as vector control methods including nets and indoor residual spraying should be used (WHO 2006). However, in areas with existing artemisinin resistance, a combination of control and elimination measures should be used to stop the survival and spread of resistance (WHO 2011).

6.10 Approaches to Fighting Resistance

To prevent or slow down the onset of resistance, the following should be done (World Health Organisation WHO. Global plan for artemisinin resistance containment (GPARC) 2011):

- Use of artemisinin combination drugs with different mechanisms of action because combination therapy slows resistance as recommended by WHO.
- Strict adherence to the dosage regimens as recommended by the manufacturers of the drugs.
- Adhering to the T3 (test, treat and track) principle of WHO. This implies testing to confirm parasite presence, treating with the right drug and tracking the patient for review.
- Definitive diagnosis of malaria before treatment, no presumptive or empirical treatment.
- Improved access to diagnostics to ensure definite diagnosis.
- Improved access to affordable, quality-assured artemisinin-based combination therapy.
- Avoidance of artemisinin monotherapy or partner drugs.
- Improve on documentation and reports of resistance.
- Increased monitoring and surveillance to evaluate resistance.

6.11 Conclusion

Resistance against antimalarial drugs has been a great public health change to the control of malaria. Transmission and spread of resistant parasites are major factors that have sustained the global burden of malaria, and this is particularly so in vulnerable groups. There is a need for continuous active surveillance and detection of molecular markers that will help in monitoring of resistance to the existing antimalarial drugs. The establishment of sentinel surveillance sites for monitoring molecular markers as well as in vivo and in vitro studies to arrest the trend of resistance will be beneficial in reducing the spread of resistant parasites.

References

Barnes KI, White NJ. Population biology and antimalarial resistance: the transmission of antimalarial drug resistance in Plasmodium falciparum. Acta Trop. 2005;94:230–40.

Bates I, Fenton C, Gruber J, Lalloo D, Lara AM, Squire SB, Theobald S, Thomson R, Tolhurst R. Vulnerability to malaria, tuberculosis, and HIV/AIDS infection and disease. Part II: determinants operating at environmental and institutional level. Lancet Infect Dis. 2004;4:368–75.

Birku Y, Mekonnen E, Bjorkman A, Wolday D. Delayed clearance of Plasmodium falciparum in patients with human immunodeficiency virus co-infection treated with artemisinin. Ethiop Med J. 2002;40(Suppl 1):17–26.

Björkman A, Bhattarai A. Public health impact of drug resistant Plasmodium falciparum malaria. Acta Trop. 2005;94(3):163–9.

Bloland PB. Drug resistance in malaria. WHO/CDS/CSR/DRS/2001.4, 2001.

Byakika-Kibwika P, Lamorde M, Mayanja-Kizza H, Merry C, Colebunders B, Van Geertruyden JP. Update on the efficacy, effectiveness and safety of artemether-lumefantrine combination therapy for treatment of uncomplicated malaria. Ther Clin Risk Manag. 2010;6:11–20.

Chijioke-Nwauche I, Wyk AV, Nwauche CA, Beshir K, Kaur H, Sutherland CJ. HIV-positive Nigerian adults harbour significantly higher serum lumefantrine levels than HIV-negative individuals seven days after treatment for *Plasmodium falciparum* infection. Antimicrob Agents Chemother. 2013;57(9):4146–50.

Chijioke-Nwauche IN. Unpublished data from PhD thesis: use of Artemether-Lumefantrine in the treatment of asymptomatic malaria in HIV-positive and HIV-negative Nigerian adults in London School of Hygiene and Tropical Medicine 2014.

Chijioke-Nwauche IN, Oguike MC, Nwauche CA, Beshir KB, Sutherland CJ. Antimalarial drug resistance markers in HIV-positive and HIV-negative adults with asymptomatic malaria infections in Port Harcourt, Nigeria. Trans R Soc Trop Med. 2021;2021(115):531–7.

Clarke SE, Jukes MC, Njagi JK, Khasakhala L, Cundill B, Otido J, Crudder C, Estambale BB, Brooker S. Effect of intermittent preventive treatment of malaria on health and education in schoolchildren: a cluster-randomised, double-blind, placebo-controlled trial. Lancet. 2008;372:127–38.

Cohen C, Karstaedt A, Frean J, Thomas J, Govender N, Prentice E, Dini L, Galpin J, Crewe-Brown H. Increased prevalence of severe malaria in HIV-infected adults in South Africa. Clin Infect Dis. 2005;41:1631–7.

Djimde A, Doumbo OK, Steketee RW, Plowe CV. Application of a molecular marker for surveillance of chloroquine-resistant falciparum malaria. Lancet. 2001;358:890–1.

Djimde AA, Doumbo OK, Traore O, Guindo AB, Kayentao K, Diourte Y, Niare-Doumbo S, Coulibaly D, Kone AK, Cissoko Y, Tekete M, Fofana B, Dicko A, Diallo DA, Wellems TE, Kwiatkowski D, Plowe CV. Clearance of drug-resistant parasites as a model for protective immunity in Plasmodium falciparum malaria. Am J Trop Med Hyg. 2003;69:558–63.

Dondorp AM, Nosten F, Yi P, Das D, Phyo AP, Tarning J, Lwin KM, Ariey F, Hanpithakpong W, Lee SJ, Ringwald P, Silamut K, Imwong M, Chotivanich K, Lim P, Herdman T, An SS, Yeung S, Singhasivanon P, Day NP, Lindegardh N, Socheat D, White NJ. Artemisinin resistance in Plasmodium falciparum malaria. N Engl J Med. 2009;361:455–67.

Duru V, Witkowski B, Menardi D. Plasmodium falciparum resistance to artemisinin derivatives and piperaquine: a major challenge for malaria elimination in Cambodia. Am J Trop Med Hyg. 2016;95(6):1228–38.

Farooq U, Mahajan RC. Drug resistance in malaria. J Vector Borne Dis. 2004;41(3–4):45–53.

Federal Ministry of Health Nigeria (FMOH). National malaria and vector control division 2005.

Handunnetti SM, Gunewardena DM, Pathirana PP, Ekanayake K, Weerasinghe S, Mendis KN. Features of recrudescent chloroquine-resistant Plasmodium falciparum infections confer a survival advantage on parasites and have implications for disease control. Trans R Soc Trop Med Hyg. 1996;90:563–7.

Hiasindh A, Subhash CP. Antimalarial drug resistance: An overview. Trop Parasitol. 2016;6(1):30–41.

Jambou R, Legrand E, Niang M, Khim N, Lim P, Volney B, Ekala MT, Bouchier C, Esterre P, Fandeur T, Mercereau-puijalon O. Resistance of plasmodium falciparum field isolates to in-vitro artemether and point mutations of the SERCA-type PfATPase6. Lancet. 2005;366:1960–3.

Jukes MC, Pinder M, Grigorenko EL, Smith HB, Walraven G, Bariau EM, Sternberg RJ, Drake LJ, Milligan P, Cheung YB, Greenwood BM, Bundy DA. Long-term impact of malaria chemoprophylaxis on cognitive abilities and educational attainment: follow-up of a controlled trial. PLoS Clin Trials. 2006;1:e19.

Laufer MK, Thesing PC, Eddington ND, Masonga R, Dzinjalamala FK, Takala SL, Taylor TE, Plowe CV. Return of chloroquine antimalarial efficacy in Malawi. N Engl J Med. 2006;355:1959–66.

Mugittu K, Ndejembi M, Malisa A, Lemnge M, Premji Z, Mwita A, Nkya W, Kataraihya J, Abdulla S, Beck HP, Mshinda H. Therapeutic efficacy of sulfadoxine-pyrimethamine and prevalence of resistance markers in Tanzania prior to revision of malaria treatment policy: Plasmodium falciparum dihydrofolate reductase and dihydropteroate synthase mutations in monitoring in vivo resistance. Am J Trop Med Hyg. 2004;71:696–702.

National Guidelines for diagnosis and treatment of malaria. Third Edition. Federal Ministry of Health National Malaria and Vector Control Division Abuja-Nigeria May 2015.

National Population Commission - NPC/Nigeria, National Malaria Control Programme - NMCP/Nigeria, and ICF International. 2012. Nigeria Malaria Indicator Survey 2010. Abuja, Nigeria: NPC/Nigeria, NMCP/Nigeria, and ICF International. http://dhsprogram.com/pubs/pdf/MIS8/MIS8.pdf.

Noedl H, Se Y, Schaecher K, Smith BL, Socheat D, Fukuda MM. Evidence of artemisinin-resistant malaria in western Cambodia. N Engl J Med. 2008;359:2619–20.

Oguike MC, Betson M, Burke M, Nolder D, Stothard JR, Kleinschmidt I, Proietti C, Bousema T, Ndounga M, Tanabe K, Ntege E, Culleton R, Sutherland CJ. Plasmodium ovale curtisi and Plasmodium ovale wallikeri circulate simultaneously in African communities. Int J Parasitol. 2011;41:677–83.

Perlmann P, Troye-Blomberg M. Malaria blood-stage infection and its control by the immune system. Folia Biol (Praha). 2000;46:210–8.

Plowe CV, Roper C, BarnwelL JW, Happi CT, Joshi HH, Mbacham W, Meshnick SR, Mugittu K, Naidoo I, Price RN, Shafer RW, Sibley CH, Sutherland CJ, Zimmerman PA, Rosenthal PJ. World Antimalarial

Resistance Network (WARN) III: molecular markers for drug resistant malaria. Malar J. 2007;6:121.

del Portillo HA, Ferrer M, Brugat T, Martin-Jaular L, Langhorne J, Lacerd MVG. The role of the spleen in malaria. Cell Microbiol. 2012;14(3):343–55.

Price RN, Cassar C, Brockman A, Duraisingh M, Van Vugt M, White NJ, Nosten F, Krishna S. The pfmdr1 gene is associated with a multidrug-resistant phenotype in Plasmodium falciparum from the western border of Thailand. Antimicrob Agents Chemother. 1999;43:2943–9.

Quinn JE, Bujila I, Cherif M, Sanou GS, Qu Y, Homann MV, Rolicka A, Sirima SB, O'Connell MA, Lennartsson A, Troye-Blomberg NI, Farrantis AO. Major transcriptional changes observed in the Fulani, an ethnic group less susceptible to malaria. elife. 2017;6(e29156):1–19.

Sawa P, Shekalaghe SA, Drakeley CJ, Sutherland CJ, Mweresa CK, Baidjoe AY, Manjurano A, Kavishe RA, Beshir KB, Yussuf RU, Omar SA, Hermsen CC, Okell L, Schallig HD, Sauerwein RW, Hallett RL, Bousema T. Malaria transmission after artemether-lumefantrine and dihydroartemisinin-piperaquine: a randomized trial. J Infect Dis. 2013;207:1637–45.

Shah SN, Smith EE, Obonyo CO, Kain KC, Bloland PB, Slutsker L, Hamel MJ. HIV immunosuppression and antimalarial efficacy: sulfadoxine-pyrimethamine for the treatment of uncomplicated malaria in HIV-infected adults in Siaya, Kenya. J Infect Dis. 2006;194:1519–28.

Sutherland CJ, Alloueche A, Curtis J, Drakeley CJ, Ord R, Duraisingh M, Greenwood BM, Pinder M, Warhurst D, Targett GA. Gambian children successfully treated with chloroquine can harbor and transmit plasmodium falciparum gametocytes carrying resistance genes. Am J Trop Med Hyg. 2002;67:578–85.

Sutherland CJ, Ord R, Dunyo S, Jawara M, Drakeley CJ, Alexander N, Coleman R, Pinder M, Walraven G, Targett GA. Reduction of malaria transmission to Anopheles mosquitoes with a six-dose regimen of co-artemether. PLoS Med. 2005;2:e92.

Sutherland CJ, Tanomsing N, Nolder D, Oguike M, Jennison C, Pukrittayakamee S, Dolecek C, Hien TT, Do Rosario VE, Arez AP, Pinto J, Michon P, Escalante AA, Nosten F, Burke M, Lee R, Blaze M, Otto TD, BarnwelL JW, Pain A, Williams J, White NJ, Day NP, Snounou G, Lockhart PJ, ChiodinI PL, Imwong M, Polley SD. Two nonrecombining sympatric forms of the human malaria parasite Plasmodium ovale occur globally. J Infect Dis. 2010;2010:1544–50.

Travassos MA, Laufer MK. Resistance to antimalarial drugs: molecular, pharmacologic, and clinical considerations. Pediatr Res. 2009;65:64R–70R.

UNDP Discussion paper: gender and malaria, December 2015.

Urdaneta L, Lal A, Barnabe C, Oury B, Goldman I, Ayala FJ, Tibayrenc M. Evidence for clonal propagation in natural isolates of Plasmodium falciparum from Venezuela. Proc Natl Acad Sci U S A. 2001;98:6725–9.

Valderramos SG, Fidock DA. Transporters involved in resistance to antimalarial drugs. Trends Pharmacol Sci. 2006;27:594–601.

Van Geertruyden JP, Mulenga M, Mwananyanda L, Chalwe V, Moerman F, Chilengi R, Kasongo W, Van Overmeir C, Dujardin JC, Colebunders RKL, D'alessandro U. HIV-1 immune suppression and antimalarial treatment outcome in Zambian adults with uncomplicated malaria. J Infect Dis. 2006;194:917–25.

Wang P, Lee CS, BayoumI R, Djimde A, Doumbo O, SwedberG G, Dao LD, Mshinda H, Tanner M, Watkins WM, Sims PF, Hyde JE. Resistance to antifolates in Plasmodium falciparum monitored by sequence analysis of dihydropteroate synthetase and dihydrofolate reductase alleles in a large number of field samples of diverse origins. Mol Biochem Parasitol. 1997;89:161–77.

White NJ. Antimalarial drug resistance. J Clin Invest. 2004;113:1084–92.

White NJ, Pongtavornpinyo W. The de novo selection of drug-resistant malaria parasites. Proc Biol Sci. 2003;270:545–54.

White NJ, Van Vugt M, Ezzet F. Clinical pharmacokinetics and pharmacodynamics and pharmacodynamics of artemether-lumefantrine. Clin Pharmacokinet. 1999;37:105–25.

World Health Organisation WHO. The use of Antimalarial Drugs: Guidance for the selection of drugs for National Antimalarial Treatment Policies. 2000.

World Health Organisation WHO. Report of Technical Consultation on Intermittent Preventive Treatment for Malaria in Infancy (IPTi), Global Malaria Programme, World Health Organization, Geneva 2006.

World Health Organisation WHO. World Malaria Report. 2011. https://www.who.int/publications/i/item/9789241564403

World Health Organisation WHO. Guidelines for the treatment of Malaria, Third Edition, 2015. https://apps.who.int/iris/bitstream/handle/10665/162441/9789241549127_eng.pdf

World Health Organisation WHO. Global plan for artemisinin resistance containment (GPARC). 2011. https://apps.who.int/iris/bitstream/handle/10665/44482/9789241500838_eng.pdf.

World Health Organisation WHO. World malaria report 2020: 20 years of global progress and challenges. https://www.who.int/publications/i/item/9789240015791

Enhancing Healthcare Through Automation and Robotics

7

Nkolika Ogechukwu Nwazor
and Stella Ifeoma Orakwue

7.1 Introduction

The mortality rate can be greatly reduced if patients have timely and easy access to quality healthcare services. This is of great importance, especially to patients in underdeveloped or developing worlds where expertise in certain fields is either not available or minimal. These patients are also faced with delays in accessing the required care because of the difficulty in manually accessing medical records, the long time taken for disease diagnosis, the lack of the requisite skills for certain procedures and the inherent human errors in these systems.

Modern technology has been of great help in overcoming some of these challenges. The introduction of information systems for easy access to medical information and records has helped make healthcare services easier and faster. Technologies like automation, robotics, the internet, artificial intelligence and the Internet of Things (IoT) are now being used in healthcare delivery. The quest for easy access to and better delivery of healthcare services has led to research into means through which these modern technologies can come to man's aid. The application of automation and robotics in the medical field was born out of the desire to use these technologies which have made man and machines do things faster and with more ease to improve healthcare and in the long run quality of life.

Automation involves the use of machines or technology to carry out tasks that would naturally have been done by human efforts while robotics involves the development of robots for specific functions. Both terms are sometimes used interchangeably although automation involves so much more. Industrial robots were initially developed to carry out tasks that are hazardous to man, access dangerous, inaccessible places, or carry out tasks requiring high precision. Currently, medical and healthcare robots are being designed for interaction with medical personnel in a surgical theatre (Fig. 7.1), assisted medical caregiving at home and medical rehabilitation (Okamura et al. 2010) in addition to the areas mentioned earlier. The resultant effects of these advancements are easy and timely access to medical attention; enhanced and timely healthcare delivery; shorter recovery times and more reliable outcomes for surgical procedures; and early disease detection, diagnosis, and treatment to mention a few. There has been tremendous growth in this sector, especially with the safety concerns raised by the Covid-19 pandemic.

Recent advancements in automation and robotics have the potential for better healthcare delivery, new treatment procedures for different ailments and health challenges, and improved

N. O. Nwazor (✉) · S. I. Orakwue
Electrical/Electronic Engineering Department, University of Port Harcourt, Port Harcourt, Nigeria
e-mail: nkolika.nwazor@uniport.edu.ng; stella.orakwue@uniport.edu.ng

E. O. Nwaichi (ed.), *Modernity in Health and Disease Diagnosis: The Account from STEM Women*, Sustainable Development Goals Series, https://doi.org/10.1007/978-3-031-34963-8_7

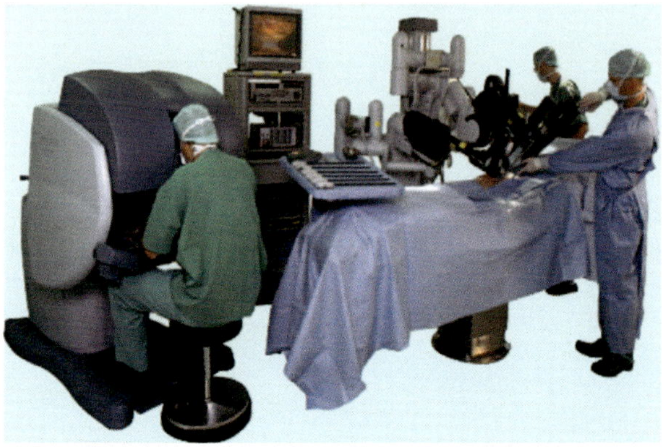

Fig. 7.1 The da Vinci surgical system consists of a master console and teleoperated patient-side robot (Okamura et al. 2010)

general patient outcomes. This work is aimed at explaining how automation and robotics have helped improve healthcare. It will also propose future areas of applications needed for improved human general well-being. The chapter starts with an introduction to automation and robotics before narrowing down to their applications in the medical field.

7.2 Overview of Automation and Robotics

The terms "automation" and "robotics" are sometimes used interchangeably, but there are minor differences between the two. In robotics human tasks, robots are used to perform specific functions while automation uses technology to perform tasks. Robots can be hardware, for example, humanoid robots, or software which are computer software applications/programs, for example, speech transcription software.

7.3 Automation

The word "automation" is derived from the Greek words "auto" translating "self" and "matos" translating "moving". Automation, therefore, is the ability for systems to "move by themselves" or operate by themselves with minimal or no human effort. However, apart from this original sense of the word, automated systems are designed to achieve significantly superior performance than what is possible with manual systems, in terms of power, precision and speed of operation.

Automation is defined as a technology concerned with the application of mechanical, electronic, and computer-based systems to monitor, operate, and control processes to increase productivity and safety, improve product quality, and reduce the cost and time of labour or execution. It focuses on the autonomous operation of systems with an emphasis on precision, efficiency, productivity, quality and reliability over an extended period. The different types of automation include fixed automation, flexible automation and programmable automation.

Automation presents a feasibility study for an entirely novel mechanism, model, or theory for applications that involve repetitive operations and most times building these models to be more efficient, reliable, or cost-effective automated systems aimed towards better use of resources, and humans to solve operating problems. (Walter 1975).

Automation can be hard or soft.

1. Hard Automation: It can be used to perform specific tasks that do not require any change in its task, for example, dishwashers, washing machines, etc.
2. Soft Automation: Here software applications are used to automate tasks that would have been performed by humans. They are flexible.

3. Autonomous Automation: The systems that fall under this make use of sensors, microprocessors and other components to make decisions. They can also have self-diagnostic and self-healing capabilities.

7.3.1 Areas of Application of Automation

1. Manufacturing: Automating a manufacturing operation such as fabrics or biscuits usually increases production rate and labour productivity. The essence of Industry 4.0 is to fully automate manufacturing processes by creating systems where every single unit is interconnected combining cloud computing and the Internet of Things to attain smart factories.
2. Logistics and supply chain management: The automation of logistics processes and ultimately the design of autonomous logistics systems is one of the most defining trends that has far-reaching consequences for the planning and execution of future logistics processes (Nitsche 2021).
3. Construction: Automation has shown the potential to increase construction productivity after years of technical development and experimenting in its field from rapid prototyping processes based on computer graphics (CAD) models to physical autonomous machines. Chena et al. (2018) describe the potential of construction automation for increasing construction productivity and the associated possible ramifications, following an objective and data-driven review of the use of automation technologies in construction.
4. Learning: Modern automation techniques such as augmented reality, in other words, creating virtual versions of devices, machines, vehicles or scenarios, have aided learning with the implementation of automated classrooms and systems.
5. Oil and gas drilling: Because oil and gas drilling often involves offshore stations and remote locations, industrial automation is extremely useful in this industry. Sensors and other monitoring equipment mean fewer difficult and potentially dangerous trips for technicians.
6. Paper mills: Automation can be used in paper mills to manage batch production, as well as control instrumentation, plant devices, and equipment. This allows operators to have excellent visibility into the entire production system.
7. Health: Automation finds great application in different areas of the healthcare sector. It can be used for disease diagnosis, patient monitoring/management, care of the elderly and patient record management to mention a few.

7.4 Robotics

Robotics uses quite many fields of technology, for example, mechanical engineering, electrical engineering, computer sciences, electronics, sensors, actuators and artificial intelligence. It is a multidimensional area that takes advantage of all engineering studies that exist in our life besides a hard mathematical module application which is required to be applied.

7.4.1 Types/Applications of Robots

1. Medical robots: They are majorly used to carry out surgical procedures. The use of robots in surgeries has greatly improved patients' outcomes and faster recovery as the surgeries are more precise and with minimal access.
2. Industrial robots: They are used in the industry for high-precision jobs and to carry out tasks that are hazardous for humans or that need to be carried out in places that are not accessible to man. Examples include painting robots, automobile assembly robots, robotic arms used for lifting heavy objects, welding robots, etc. They are flexible and are not fatigued since their repetition rate is high.

3. Domestic robots: These robots are used to carry out household chores like switching on and off electrical and electronic appliances in the home, sweeping the house, opening and shutting doors, house cleaning, plate washing, and others.
4. Mobile robots: They can move around and are mainly used to access hazardous areas like chemical or nuclear power plants. They are also used in obstacle detection and avoidance.
5. Unmanned aerial vehicles and boats: These are aircraft and boats without a pilot. They are mainly used for rescue missions or attacks on enemy territories. They are also used for surveillance.
6. Military and security robots: Robots find applications in the military in the areas of missile launching and aerial surveillance. Some companies are also working on pairing robots with human security guards for enhanced security.
7. Humanoids: These are robots that take human-like forms and can perform functions that ordinarily would have been performed by humans.
8. Cobots: They are robots that function with humans directly or alongside humans. They can share spaces with humans to help them perform more tasks (Intel 2022). They sometimes mimic human actions.
9. Agribots: They are used in agriculture, especially to take care of repetitive and hazardous tasks like weed control, seeding, harvesting, chemical applications, food processing and irrigation, for example, Ecorobotix (GeekforGeek 2020).

7.4.2 Types of Medical Robots (Intel 2020a)

Medical robots can be grouped into four:

1. Autonomous mobile robots (AMR): They are robots that can understand and move through their environment easily. They find applications in tasks like the delivery of medical supplies and medication and hospital disinfection. They can also be used to take patients' basic parameters thereby reducing the workload of the hospital staff while ensuring their safety.
2. Social robots: They are interactive robots used mainly in long-term care environments to monitor patients and also interact with them. They can provide cognitive engagement, can encourage patients to comply with their treatment regimen, and can serve as guides to both patients and visitors. They help improve patients' emotional health while reducing the burden of care.
3. Surgical-assistance robots: They are robots that help surgeons in the operating room to achieve more precision, and accuracy and work with improved speed during complex operating procedures. Some surgical robots are autonomous (they perform the surgeries while the surgeons supervise remotely). The surgeries can be minimally invasive like hysterectomy, bariatric surgery and other soft tissue surgeries. With the help of surgical robots, infections and complications are greatly reduced. The surgeries can also be orthopaedic like knee replacements, hip replacements, etc. With videoconferencing capabilities, surgeons can consult with others in different locations. With this, patients can get the best surgeons involved in their procedures. They help surgeons reduce nerve damage during surgeries. Soon, robots will be able to suture incisions. Surgical robot training for surgeons through simulation platforms can be carried out using simulation platforms.
4. Modular robots: They perform multiple functions and are used to enhance other systems, for example, prosthetic robotic arms and therapeutic robots for rehabilitation. A good example is the wheelchair-mounted robotic arm (Fig. 7.2) that helps patients with spinal injuries with their daily activities. They can also be used to monitor patients as they go through exercises or other rehabilitation procedures.

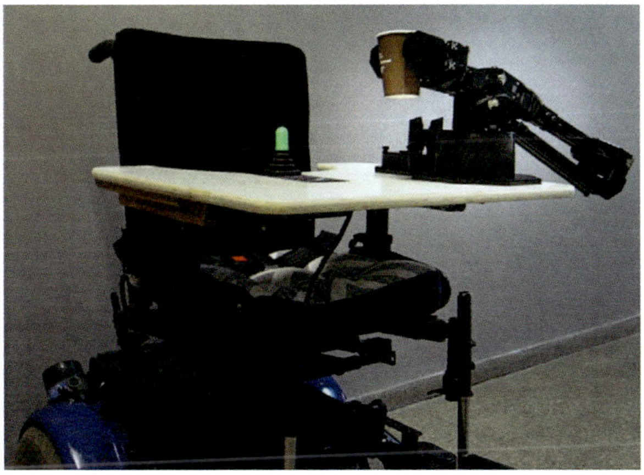

Fig. 7.2 Intel and Accenture robotic arm by Neuromorphic Research Project to assist wheelchair-bound paediatric patients (Intel 2020b)

7.5 Automation and Robotics in Healthcare

7.5.1 Applications of Automation in Healthcare

In this era of serious economic crisis when healthcare administrators are saddled with the responsibility of enhancing the quality of care received by patients, reducing costs and supporting decision-making using data analysis, automation comes in readily as a major consideration. Automation finds applications in different aspects of healthcare. It can be used in administration, disease diagnosis and treatment, medical training and research to mention a few. In pursuance of SDG 3, the areas of application of automation in the healthcare industry include:

1. Patients' and doctors' appointment scheduling: Software applications are used to automate patients' appointments with doctors depending on the needs and the doctors' availability.
2. Financial management: All financial transactions in the healthcare industry can be automated. Patient billing, payments, salaries/claims by staff, expenditures and other revenue generation can be taken care of using software applications.
3. Healthcare staff protection/safety: Patients' contact with caregivers can be minimized using automation. Tasks that would have required interviews by a nurse or other healthcare professionals can be automated to enable self-services for patients. The advent of COVID-19 and other pandemics have led to the invention of mechanisms that enable patients to self-triage and feed in their information for access by professionals in protected spaces. Contacts are only allowed where necessary. Other tools have been developed to diagnose some ailments without direct contact with caregivers by observing defined parameters. This also goes a long way in reducing the amount of work done by caregivers.
4. Medical record keeping: This involves storing patients' records in electronic format. The collection, processing, storage and retrieval of the large amount of data generated in the medical records department can be automated. This will aid in the timely retrieval of patients' records when needed by the caregivers. It will also enhance the quality of care as patients' records can be accessed anytime from anywhere using the internet and centralized databases. This will also aid research as data can easily be collated.
5. Diagnosis: A lot of applications exist online that are used for diagnosis. The patient can interact with the system through a user-friendly graphical user interface, and feed in

their information as responses to predefined queries on the platform which the system analyses to diagnose the patient's health challenge and make a recommendation on treatments or schedule the patient for an appointment with a caregiver. Automation has greatly enhanced clinical diagnosis and decision-making. It supports evidence-based medical practice to identify best practices in disease diagnosis and treatment.

6. Surgical procedures: Surgical procedures are now being carried out by professionals in remote locations with the aid of experts that are not onsite. Augmented reality, which makes use of 3D modelling, has been used to aid this process.
7. Smart devices: The use of smart devices (hard automation) to monitor critical parameters like blood pressure, pulse, temperature, heart rate, etc., has greatly improved patient outcomes. Some of these smart devices make use of wearable devices. These devices can monitor patients' parameters, diagnose patients' ailments timely and even recommend treatments and preventive measures that can be taken by patients. With the introduction of the Internet of Things, some of these devices are also connected to other systems in the healthcare facility and can even communicate with the patient's caregiver for timely intervention in cases of emergency and effective monitoring.

7.5.2 Application of Robotics in Healthcare

Areas of applications of robotics in the healthcare industry include:

1. Communicable and infectious diseases with high mortality rate: To protect healthcare workers in situations where you have highly contagious diseases with high mortality like Ebola, COVID-19, etc., robots can be used instead of the human caregiver. To reduce physical contact between patients and caregivers, robotics and automation are needed in health facilities (Yang et al. 2020a; Yang et al. 2020b) for routine activities like patient monitoring, drug dispensing, food delivery, facilities disinfection, etc. This will reduce the cost of care as resources needed for procuring protective gear can be channelled to other areas of need.
2. Surgery: Robots are being used to carry out surgical procedures. A robot called NeuroArm was used with real-time magnet resonance imaging for image-guided neurosurgery to remove a brain tumour (Bogue 2011). It has also been used for deep brainstem stimulation and catheter implantation.
3. Prostheses: Robotics has been implemented in the development of artificial hands and legs for amputees (Bogue 2011). Grippers have also been developed (Dollar and Howe, 2010). Robots have been developed for wheelchair-bound patients to enable them to stand and turn at certain predefined angles.
4. Patient care: Robots are used in patient monitoring and support (Kuka 2023). It has been applied in both mental and physical support as reminders, for emotional support, education of children with mental disabilities, drug and food delivery, etc. (Kyrarini et al. 2021). Some of the patient care robots developed include Pepper used as an assistant to the elderly and children (Tanaka et al. 2015; Yang et al. 2017) and support to psychiatric patients (Sato et al. 2020); Care-O-bot used as a caregiver robot in a care facility for the elderly (Jacobs and Graf 2012); and PHAROS which is for monitoring the daily physical activities of the elderly in their homes (Martinez-Martin et al. 2019) to mention a few. The advancements in technology-based healthcare gave birth to Healthcare 4.0 which led to the introduction of cyber-physical systems-based homecare robotic systems (CPS-HRS) which are faster and more intelligent as proposed by Yang et al. (2020a, 2020b). CPS-HRS incorporates motion capture and mapping and uses artificial intelligence techniques, advanced sensor technologies and other current technologies to monitor behaviour, especially of the elderly in a natural setting like home (Portet et al. 2013).

This can be applied to monitoring the progression of a disease, detecting and preventing falls (Kau and Chen 2015), etc. They can also be used to carry out some medical tests.

5. Diagnosis and treatment: Robotic systems are being employed in disease diagnosis as a safety measure to reduce contact between doctors and patients. It has been used to perform medical tests (Nemati et al. 2012; Majumder et al. 2017a, 2017b).
6. Robotic exoskeletons: They act as an external set of muscles and bones that help train the body on how to move properly (Banks 2022). They are used for rehabilitation as they can help disabled people regain mobility. With these robots, immobile patients can move around, which has emotional, psychological and physical therapeutic effects. With these, patients are made to have a feeling of some level of independence which has a booster effect on the recovery process.

7.6 Challenges/Ethical Issues of Medical Automation and Robotics

There have been several arguments concerning ethics on the use of robots and automated systems in healthcare. There is the challenge of its impact on human relationships which has been observed to positively impact patient recovery. Empathy and compassion are human qualities that significantly affect healing, especially in attending to paediatric, gynaecological, and psychiatric patients, and are lacking in these systems. There is fear that robots may dehumanize medical care as they are not empathic and compassionate (Stahl et al. 2014; Stahl and Coeckelbergh 2016).

Dependence on robots for surgical procedures may lead to loss of surgical skills overtime as surgeons depend more on the robots for the performance of procedures that they ordinarily would have carried out and concentrate more on developing skills on how to manipulate and control the robots (Saniotis and Henneberg 2021). The study by Kyrarini et al. (2021) shows that most patients prefer human caregivers to robots and would not want a total replacement. They also complained about the difficulty in understanding some of the actions of the robot and the challenges of the malfunctioning of the robot and who should be held responsible if the system fails. Liu et al. (2013) talked about privacy and security challenges concerning the healthcare data generated by these systems. Patients' data collected through automation and robotics can be hacked into and used maliciously (Farhud and Zokaei 2021). Another major challenge in the use of medical robots and automation is its effect on employment. Over-dependence on robots for the performance of jobs done by a human may eventually lead to job loss in the nearest future.

Other challenges in the application of automation and robotics in healthcare include training the healthcare workers on the use of the new systems, and improvement of the operations of the robots which requires more research as they cannot be said to be completely error-free.

7.7 The Future of Medical Automation and Robotics

The next generation of healthcare systems will be greatly powered by automation and robotics due to their technological advancements. They will be of great help in assessing war zones for delivery of medical aid to those at the war front, managing highly infectious diseases like the Ebola disease outbreak and during pandemics as witnessed during the COVID-19 pandemic. They can aid in sample collection, caregiving to affected/infected patients, supply logistics, hospital disinfection and patient monitoring and rescue.

The introduction of the Internet of Things and specifically the Internet of Healthcare Things, IoHT (Kaiser et al. 2021), that connect patients; healthcare staff (nurses, physicians, caregivers, patients and other hospital support staff); medical infrastructure (like ambulances, laboratories); and medical instruments/gadgets both in the hospital and patient's home will greatly enhance the quality of care and reduce human capital costs in such situations (Kaiser et al. 2021).

A combination of artificial intelligence, robotics and automation will help increase precision and accuracy in medical procedures. The autonomous robotic ultrasound was found to be more effective during transcranial Doppler (TCD) in discovering serious cardiac issues that were completely missed using a standard of care imaging (Hamilton 2022).

Blockchain technology which is a decentralized core architecture that adopts distributed accounting, communication and storage with the integration of technology can be used to enhance the security of data generated by automated and robotic systems since it can be used to give users different levels of access to the generated data (Zyskind and Nathan 2015; Azaria et al. 2016). Further research is encouraged on the application of blockchain technology and other security systems in securing healthcare-generated data. Big data analytics can also be applied to predict the occurrence of health events in patients even before they occur (Majumder et al. 2017b; Naghshvarianjahromi et al. 2019). The advancements in technologies like data analytics, artificial intelligence, machine learning, and computer vision will help enhance the functionality of medical robots as they carry out tasks more efficiently, accurately and autonomously. More research is encouraged to greatly improve the efficiency of these automated and robotic systems to make them safe and good enough to be used as an alternative to the human-driven healthcare system. Research into new areas of applications is also encouraged.

In the nearest future, it is believed that more improvements in the development of microbots (Banks 2022), which are tiny robots that can navigate through the human body, will greatly reduce recovery time from surgical procedures. With more enhancements, these robots can even do surgery from within the body, removing the need for surgical incisions. Research into body-healing microbots which can be as tiny as the body cells are greatly encouraged.

7.8 Conclusion

The introduction of automation and robotics in the healthcare industry has greatly improved medical outcomes the challenges notwithstanding. It has made access to medical care easier and faster. It has greatly reduced recovery time and made the healing process safer, cheaper, smarter, and less stressful for both healthcare workers and patients. It has also improved caregiving/medical outcomes and ultimately improves life. The capability is not exhaustive and therefore more research is encouraged in this area. The ethical issues raised in this work should also be taken into consideration in the use of these systems.

References

Azaria A, et al. Medrec: using blockchain for medical data access and permission management. In: Proc. int. conf. Open big data, Germany; 2016. p. 25–30.

Banks, M. How robots are redefining health care: 6 recent innovations. robotics tomorrow; (2022). December 22, 2022 from https://www.roboticstomorrow.com/story/2022/03/how-robots-are-redefining-health-care-6-recent-innovations/18339/

Bogue R. Robots in healthcare. Ind Robot Int J. 2011;38(3):218–23. https://doi.org/10.1108/01439911111122699.

Chena Q, de Sotob BG, Adey BT. Construction automation: research areas, industry concerns and suggestions for advancement. Autom Constr. 2018;94:22–38. https://doi.org/10.1016/j.autcon.2018.05.028.

Dollar A, Howe R. The Highly Adaptive SDM Hand: Design and Performance Evaluation. International Journal of Robotic Research. 2010;29:585–97. https://doi.org/10.1177/0278364909360852.

Farhud DD, Zokaei S. Ethical issues of artificial intelligence in medicine and healthcare. Iran J Public Health. 2021;50(11):v. https://doi.org/10.18502/ijph.v50i11.7600.

GeeksforGeeks. Top 10 Applications of Robotics in 2020. 2020. December 10, 2022 from https://www.geeksforgeeks.org/top-10-applications-of-robotics-in-2020/

Hamilton, R. Robotics and automation are the future of healthcare. Medcity influencers. 2022. Retrieved December 15, 2022 from https://medcitynews.com/2022/08/robotics-and-automation-are-the-future-of-healthcare/

Intel. Robotics in Healthcare: The Future of Robots in Medicine. 2020a. Retrieved December 15, 2022 https://www.intel.com/content/www/us/en/healthcare-it/robotics-in-healthcare.html

Intel. Intel and Accenture Support Neuromorphic Research Project to Assist Wheelchair-Bound Pediatric Patients,

Intel Newsroom. 2020b. Retrieved December 15, 2022 from https://www.intel.com/content/www/us/en/newsroom/news/neuromorphic-research-wheelchair-pediatric.html#gs.ogik2n

Intel. Types of Robots: How Robotics Technologies Are Shaping Today's World. (2022). Retrieved December 23, 2022 from https://www.intel.com/content/www/us/en/robotics/types-and-applications.html

Jacobs T, Graf B. Practical evaluation of service robots for support and routine tasks in an elderly care facility. In: Proceedings of the 2012 IEEE workshop on advanced robotics and its social impacts (ARSO); 2012.

Kaiser MS, Al Mamun S, Mahmud M, Tania MH. Healthcare robots to combat COVID-19. In: Santosh KC, Joshi A, editors. COVID-19: prediction, decision-making, and its impacts, lecture notes on data engineering and communications technologies. Springer Nature Singapore Pte Ltd; 2021. p. 60. https://doi.org/10.1007/978-981-15-9682-7_10.

Kau LJ, Chen CS. A smart phone based pocket fall accident detection, positioning, and rescue system. IEEE J Biomed HealthInf. 2015; 19(1):44–56.

Kuka (2023) Pooling expertise with a goal towards automating the medical sector. https://www.kuka.com/en-us/industries/health-care

Kyrarini M, Lygerakis F, Rajavenkatanarayanan A, Sevastopoulos C, Nambiappan HR, Chaitanya KK, Babu AR, Mathew J, Makedon FA. Survey of robots in healthcare. Technologies. 2021;9(1):8. https://doi.org/10.3390/technologies9010008.

Liu, et al. Completely pinpointing the missing RFID tags in a time-efficient way. IEEE Trans Comput. 2013;64(1):87–96.

Majumder S, Aghayi E, Noferesti M, Memarzadeh-Tehran H, Mondal T, Pang Z, Deen MJ. Smart homes for elderly healthcare—recent advances and research challenges. Sensors. 2017b;17(11):2496.

Majumder S, Mondal T, Deen MJ. Wearable sensors for remote health monitoring. Sensors. 2017a;17(1):130.

Martinez Martin E, Costa A, Cazorla M. PHAROS 2.0—a PHysical assistant robot system improved. Sensors. 2019;19:4531

Naghshvarianjahromi M, Kumar S, Deen MJ. Brain-inspired intelligence for real-time health situation understanding in smart e-health home applications. IEEE Access. 2019,7:180106–26.

Nemati E, Deen MJ, Mondal T. A wireless wearable ECG sensor for long-term applications. IEEE Commun Mag. 2012;50(1):36–43.

Nitsche B. Exploring the Potentials of Automation in Logistics and Supply Chain Management: Paving the Way for Autonomous Supply Chains. Logistics. 2021; 5(3):51. https://doi.org/10.3390/logistics503005.

Okamura AM, Mataric MJ, Christensen HI. Medical and healthcare robotics: achievements and opportunities. IEEE Robot Autom Mag. 2010:26–37. https://doi.org/10.1109/MRA.2010.937861. https://www.researchgate.net/publication/224173784_Medical_and_Health-Care_Robotics/figures?lo=1

Portet F, Vacher M, Golanski C, Roux C, Meillon B. Design and evaluation of a smart home voice interface for the elderly: acceptability and objection aspects. Pers Ubiquit Comput. 2013;17(1):127–44.

Saniotis A, Henneberg M. Neurosurgical robots and ethical challenges to medicine. Ethics Sci Environ Politics. 2021;21:25–30. https://doi.org/10.3354/esep00197.

Sato M, Yasuhara Y, Osaka K, Ito H, Dino MJS, Ong IL, Zhao Y, Tanioka T. Rehabilitation care with pepper humanoid robot: a qualitative case study of older patients with schizophrenia and/or dementia in Japan. Enferm Clín. 2020;30:32–6.

Stahl BC, Coeckelbergh M. Ethics of healthcare robotics: towards responsible research and innovation. Robot Auton Syst. 2016;86:152–61. https://doi.org/10.1016/j.robot.2016.08.018.

Stahl BC, McBride N, Wakunuma K, Flick C. The empathic care robot: a prototype of responsible research and innovation. Technol Forecast Soc Change. 2014;84:74–85. https://doi.org/10.1016/j.techfore.2013.08.001.

Tanaka F, Isshiki K, Takahashi F, Uekusa M, Sei R, Hayashi K. Pepper learns together with children: development of an educational application. In: Proceedings of the 2015 IEEE-RAS 15th international conference on humanoid robots (humanoids), Seoul, Korea, 3–5 November 2015; 2015. p. 270–5.

Walter R. A concept of factory automation. IEEE Trans Manufact Technol. 1975;4(2):56–8. https://doi.org/10.1109/tmft.1975.1135865.

Yang C, Lu M, Tseng S, Fu L. A companion robot for daily care of elders based on homeostasis. In: Proceedings of the 56th annual conference of the Society of Instrument and Control Engineers of Japan (SICE), Kanazawa, Japan, 19–22 September 2017; 2017. p. 1401–6.

Yang G, Pang Z, Deen MJ, Dong M, Zhang YT, Lovell N, Rahmani AM. Homecare robotic Systems for Healthcare 4.0: visions and enabling technologies. IEEE J Biomed Health Informatics, (Early Access). 2020a; https://doi.org/10.1109/jbhi.2020.2990529.

Yang GZ, Nelson BJ, Murphy RR, et al. Combating COVID-19—the role of robotics in managing public health and infectious diseases. Sci Robotics. 2020b;5(40) https://doi.org/10.1126/scirobotics.abb5589.

Zyskind G, Nathan O. Decentralizing privacy: using blockchain to protect personal data. In: Proc. IEEE Secur. Priv. Workshops; 2015. p. 180–4.

8 Naturally Derived Surfactants for Healthy Food Formulation

Millicent Uzoamaka Ibezim-Ezeani

8.1 Introduction

With the exponentials increase in human population worldwide, it is expected that global food demand will rise significantly and so the formulation of foods that are compatible with the body system is necessary in order to achieve food security and improved nutrition for the benefit of mankind, as well as food preservation and storage, while ensuring sufficiency of healthy food all round seasons of the year (Sustainable Development Goals—1, 2, 3, and 12) (Morton et al. 2017).

Food is an edible substance composed mainly of carbohydrates, proteins, fats, lipids, etc., which when consumed by living organisms provides the required nourishment for the sustenance of its existence. According to Nwaichi and Ntorgbo (2016), human health is largely determined by the diet and a recommendable diet should be able to provide sufficient nutrients and with tolerable levels of pathogenic microorganisms and chemical contaminants. Thus, formulated foods are those food materials prepared with ingredients which boost the supplemental influence of minerals, vitamins, and condiments to the body system on ingestion.

M. U. Ibezim-Ezeani (✉)
Department of Pure and Industrial Chemistry, University of Port Harcourt, Choba, Rivers State, Nigeria
e-mail: millicent.ibezim-ezeani@uniport.edu.ng

In addition to the nutriments gained from eating formulated foods, they furnish the body with nutraceutical benefits which in many cases support vascular function, maintain a healthy brain, enhance systemic circulation, rejuvenate body organs, promote muscular development, make up for dietary deficiencies, ensure vitality, energize body system interactions, and so on.

Food formulation, therefore, involves an assembled collection of microstructures of carbohydrates, proteins, fats, lipids, etc., stabilized with naturally derived surfactants mainly through the process of emulsification, solubilization, dispersion, foam formations, etc., which are specifically tailored to the nutritional needs of humans and other organisms. Some examples of formulated foods are margarine, oat, baking and wheat flour, mayonnaise, desserts, bread, confectioneries, tinned corn, canned tomatoes, etc. (Sharma 2014).

8.2 Naturally Derived Surfactants

Naturally derived surfactants (NDSs) designed for healthy food formulation are surface active-stabilizing molecules (of organic origin) with functionalized amphipathic efficacy at the surface or interface which are coherently free from artificial chemicals and certified by recognizable regulatory agencies. These functionalized NDSs

E. O. Nwaichi (ed.), *Modernity in Health and Disease Diagnosis: The Account from STEM Women*, Sustainable Development Goals Series, https://doi.org/10.1007/978-3-031-34963-8_8

could be applied as separable components (extracts/filtrates/exudates/isolates/distillates/concentrates/fermentates) or derivatives with edible-grade solvents. Some of the functional constituents inherent in parts of plants or animals which have contributed to the stability of complex colloids and emulsions in formulated foods are lecithin, glutens, amino acids, leucine, alkaloids, alanine, steroids, tannins, glycosides, lipids, phenolics, flavonoids, and aspartic-, glutamic-, and citric-acids (Holmberg 2001; Chhetri et al. 2009; Hosseinzadeh et al. 2013; Atta et al. 2021; Mehrjoo et al. 2022). These innate attributes of the dual-structured character of NDSs contribute to their wholesome interactions in the food formulation process as emulsifying, lubricating, stabilizing, shelf life–extending, dispersing, wetting, spreading, crystal-modifying, fermenting, conditioning, dough gas–retaining, strengthening, crystallization-inhibiting, aerating, stick-preventing, quality-improving, foaming/antifoaming, flavor-enhancing, gel-forming, and moisture-retaining agents or additives (Sharma 2014).

They acquire thermodynamic and kinetic stabilities through their preferential orientation at the surface or interface of colloidal systems, with the polar (hydrophilic, lipophobic, or oleophobic) head portion and non-polar (hydrophobic, lipophilic, or oleophilic) tail portion (Fig. 8.1) aligning towards aqueous and organic phases, respectively.

8.3 Classification of Naturally Derived Surfactants

Naturally derived surfactants are classified based on the capacity of their hydrophilic head groups to exercise influential attraction for ionic moieties in colloidal fluid. The four major classes of NDSs (Fig. 8.2) are the naturally derived cationic surfactant, naturally derived anionic surfactant, naturally derived nonionic surfactant, and naturally derived zwitterionic surfactant (Schramm et al. 2003; Massarweh and Abushaikha 2020; Atta et al. 2021; Isaac et al. 2022).

8.3.1 Naturally Derived Cationic Surfactant

Naturally derived cationic surfactant (NDCS) carries a positively charged head group with

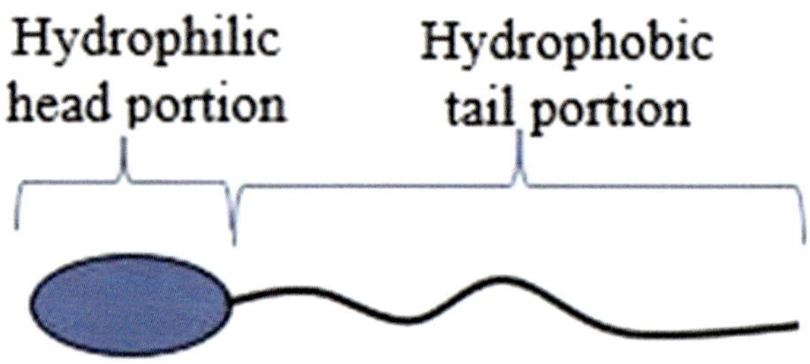

Fig. 8.1 Illustration of an NDS molecule

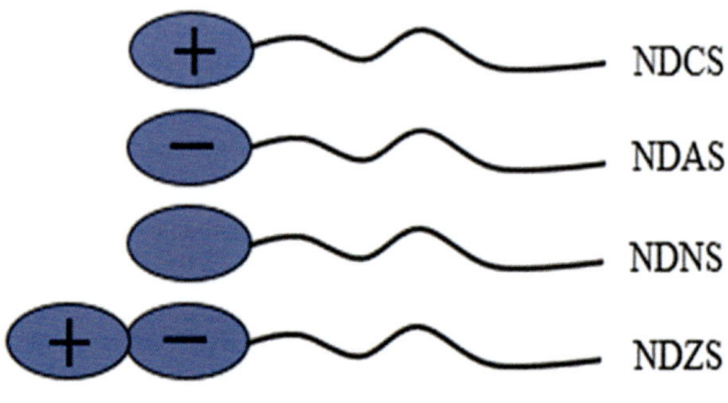

Fig. 8.2 Schematic representation of classes of NDSs

compositions such as ammonium salts of fatty acids, fatty diamine salts, diester amine quaternary salts, and simple fatty amine salts.

8.3.2 Naturally Derived Anionic Surfactant

Naturally derived anionic surfactant (NDAS) bears a negatively charged head group with components made up of carboxylates, sulfates, sulfonates, and phosphates.

8.3.3 Naturally Derived Nonionic Surfactant

Naturally derived nonionic surfactant (NDNS) possesses zero net charged head group with common constituents such as alkylphenol ethoxylates, poly-propoxylated alcohols, and poly-ethoxylated alcohols.

8.3.4 Naturally Derived Zwitterionic Surfactant

Naturally derived zwitterionic surfactant (NDZS) has both positively and negatively charged head groups, with the cationic part of the hydrophile serving as the link to the hydrophobic chain. Acids of sulfonate, phosphate ester, carboxylate, and sulfate groups are instances of a charged center for the anionic part, while amines and quaternary ammonium are examples of functional group components that make up the charged center for the cationic part of NDZS.

For instance, as part of the structural activity of NDSs in the food formulation process (Sharma 2014):

(a) Arginine and lysine exhibit their cationic feature through the influence of the amino group-head portion at the interface of colloidal dispersion.
(b) Aspartic and glutamic acids contain carboxyl groups at the side chains which enhance their anionic behavior in colloidal solution at the interface.
(c) Alanine and leucine possess aliphatic side chains with non-polar hydrophile-controlled character at the interface in the fluidous colloidal system.

8.4 Characterization of Naturally Derived Surfactants

With the adsorption of NDSs, the interfacial (surface) tension and free energy of the system are relatively reduced by proportions which are connected to the added surfactant quantity; thus, the unification of the two immiscible aqueous and oily phases start at the attainment of a particular concentration to form micelles (micellization).

Naturally derived surfactant molecules are characterized based on their:

(a) Active influence at the surface or interface of the system.
(b) Polar head and nonpolar tail structural frame of the moiety.
(c) Self-organized association between the amphiphilic components.

The principles governing food emulsions, micelle formation, hydrophilic-lipophilic balance, and inter-micellar interactions are the main characteristic considerations before the application of NDSs for healthy food formulation.

8.4.1 Emulsions in Food Formulation

The prolonging of shelf life under the condition of enduring stability of formulated food structural forms is dependent on the thorough grasp of the physicochemical characteristics and transformational behavior of emulsions in foods, as well as its implementation in the developmental pattern of consumable-type NDSs for food formulation. The dispersion of an immiscible liquid (dispersion phase) into another liquid (dispersion medium or continuous phase) with which it is stabilized by a suitable food-grade NDS is referred to as emulsion in food formulation. Oil-in-Water (O/W), Water-in-Oil (W/O), and Water-in-Oil-in-Water

(W/O/W) emulsions are the prevalent type of emulsions governing the food formulation process (Fig. 8.3) (Schramm et al. 2003; Kralova and Sjöblom 2009; Sharma 2014).

8.4.1.1 Oil-in-Water Emulsion

This is a colloidal system which involves an oily phase dispersed in a continuous phase of water. The applied NDSs and any impurity present in the water phase dictate the physicochemical behavior of the emulsion. Some examples of O/W emulsion in formulated foods are creamers, cream liqueur, whipped toppings, ice cream mixes, and mayonnaise.

8.4.1.2 Water-in-Oil Emulsion

This system is a colloidal suspension of droplets of water (dispersion phase) in an insoluble mixture with an oily dispersion medium. The physicochemical properties of the NDSs and the purity level of the lipids used determine the stability of the W/O emulsion. Some examples of W/O emulsions in formulated foods include butter, margarine, and fat-based spreads.

8.4.1.3 Water-in-Oil-in-Water Emulsion

This is a colloidal system made up of an oil phase suspended an in aqueous phase, with the oil phase having small droplets of water dispersed in it. In order to have a significantly stable W/O/W emulsion, the NDS, water, and oil used should be free of foreign materials. Many bakery products belong to this class of emulsion.

8.4.2 Micellization of Naturally Derived Surfactants

Micellization is a key behavior of all surfactants, and the concentration of surfactants above which micelles are formed is termed the "critical micelle concentration" (CMC) (De et al. 2015;

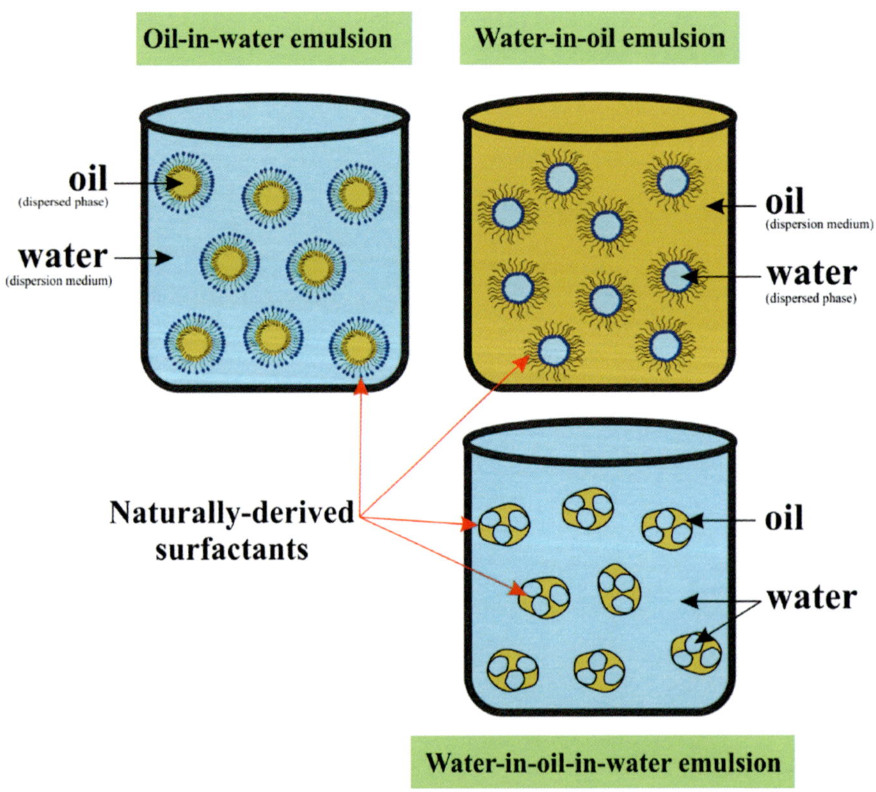

Fig. 8.3 Types of emulsions in food formulation

Massarweh and Abushaikha 2020; Mehrjoo et al. 2022). The measurements of surface or interfacial tension plotted as a function of surfactant concentrations are often applied in evaluating the CMC of NDSs. The concentration of surfactant which coincides with the breakoff point from the slopy-line plot depicts the CMC of that particular NDSs (Fig. 8.4).

The pre-micellar region in Fig. 8.4 denotes the interval of preferential orientation of surfactant molecules at the interface with an increase in its quantity, followed by the attendant degree of reduction of interfacial tension in this circumstance. Thus, with further increase and as time goes on, the saturation of surfactant concentration at the ınterconnection point of the phases is reached when the self-generation of micelles begins to occur which is depicted as CMC (0.034 mg/L). The post-CMC (plateau) region signifies the period in which more micelles are formed with further addition of surfactant molecules; followed by an abrupt change in the physicochemical properties with relatively no noticeable variation in parameters such as electrical conductivity, thermal resistivity, and interfacial/surface tension. This shows that they do not rely on an increase in the concentration of surfactants any longer, but that extra surfactant molecules self-organize at a specific quantity to form more micelles.

8.4.3 Hydrophilic-Lipophilic Balance

The ability to generate suspension between two immiscible liquids and how the stabilizing magnitude of NDS molecules influences the behavior of hydrophilic and lipophilic moieties are made known by the hydrophilic-lipophilic balance (HLB) value. The HLB values are scaled from 0 to 20, which reveals the spontaneous tendency towards oil or water emulsion (Kralova and Sjöblom 2009). Thus, a low value of HLB points to NDS's high capacity for oil emulsion, while a high value of HLB is an indication of NDS's high capacity for water emulsion. Table 8.1 shows the range of HLB values for emulsifying capabilities of NDS applications (Massarweh and Abushaikha 2020).

Thus, the HLB values of NDSs (separable components, intermixtures, or formulations) can

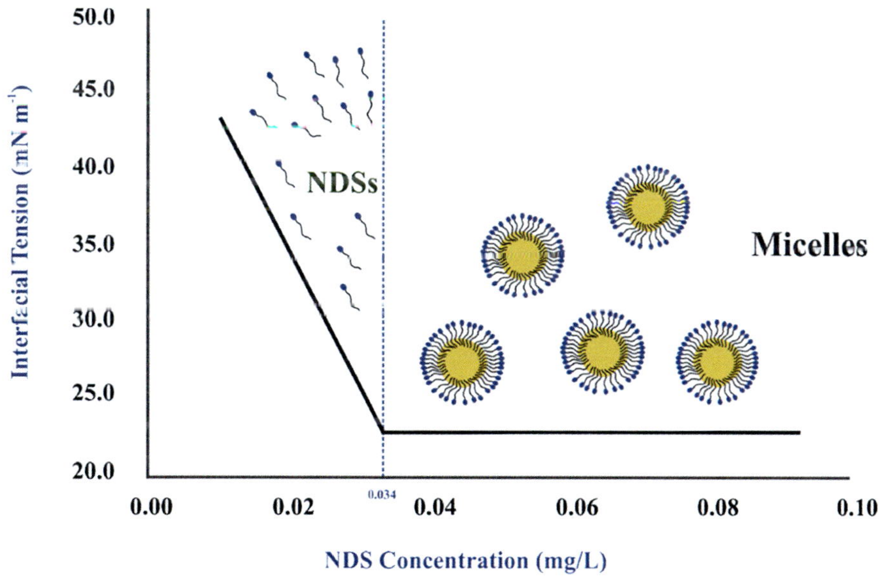

Fig. 8.4 Plot of interfacial tension versus NDS concentration

Table 8.1 Prediction of HLB values with NDS applications

HLB value	Application
0–3	Antifoaming agent
4–6	Water-in-oil emulsifier
7–9	Wetting agent
8–18	Oil-in-water emulsifier
13–15	Detergent
10–18	Solubilizer

be evaluated experimentally using the formula in Eq. 8.1:

$$\mathrm{HLB} = 20\left(1 - \frac{S}{A}\right) \quad (8.1)$$

where S and A are the saponification and acid values, respectively.

8.4.4 Inter-Micellar Interactions of Naturally Derived Surfactants

In the inter-particle interactions of colloidal fluids, the capacity of the tail moieties to produce a hydrophobic effect and the power of electrostatic or steric repulsion exerted between the hydrophilic portions are the main contending forces that control the micellar aggregation and size, respectively; while the hydrophobic and hydrophilic groups of the individual surfactant particle determine the shape of the micelle (Goyal and Aswal 2001). In order to group micellar aggregation intelligibly along a systematic sequence, the idea of molecular packing parameter advanced by Israelachvili et al. (1977) was generally applied to describe the implication of the geometric relationship between the tail (volume and length) and head (area) features to the self-organization of surfactant molecules in fluidous colloids.

Research scientists and industrial technologists in the field of food processing and formulation have within the past two and half decades applied the molecular packing parameter as a means of providing uncomplicated and perceptive insight into the multimolecular assemblage of structured shapes (spherical, cylindrical, vesicles, planar bilayers, and reverse) formed in colloidal solutions (Fig. 8.5), which have impacted value over rheological property, emulsification capacity, solubilization performance, and dispersion capability of NDSs in the food formulation industry.

The molecular packing parameter (P) is expressed as (Nagarajan 2002; Stuart and Boekema 2007; Yan et al. 2007; Massarweh and Abushaikha 2020):

$$P = \frac{v}{a_e l} \quad (8.2)$$

where v and l are the surfactant's tail volume and length, respectively, while a_e is the equilibrium area per molecule at the aggregate interface.

The v and l parts of the expression are consistent variables for a particular NDS molecule, and since they are determined from the innate features of the molecular structure, the ratio of v to l will remain constant. However, the surface area is a product of compilation from the circumference, radius, and height of the formed structure due to inter-micellar interactions, which is dependent on the extent of surface accessed by the head group of NDS molecule at the interface of the hydrophobic core-hydrophilic media under equilibrium condition. Thus, modifications in micellar solution conditions (such as temperature, pH, ionic strength, and concentration) will likely alter the value of P through the influence over the surface area variable and, in turn, lead to the aggregate morphology transition.

For instance, the salt added to ionic micellar solution shields the colloidal fluid from the repulsive force between two positive or two negative charges and, hence, allows the modification of surface area which will alter the P value. Thus, transform the inter-micellar interactions to the aggregate morphology that yields the desired products with properties like longer shelf-life and improved quality. Furthermore, variation in micellar solution to a higher temperature can moderate the performance of non-ionic NDSs to a new value of the corresponding P due to hydrogen bonding interaction at the interface; thereby achieving the product of the required structural form (Goyal and Aswal 2001).

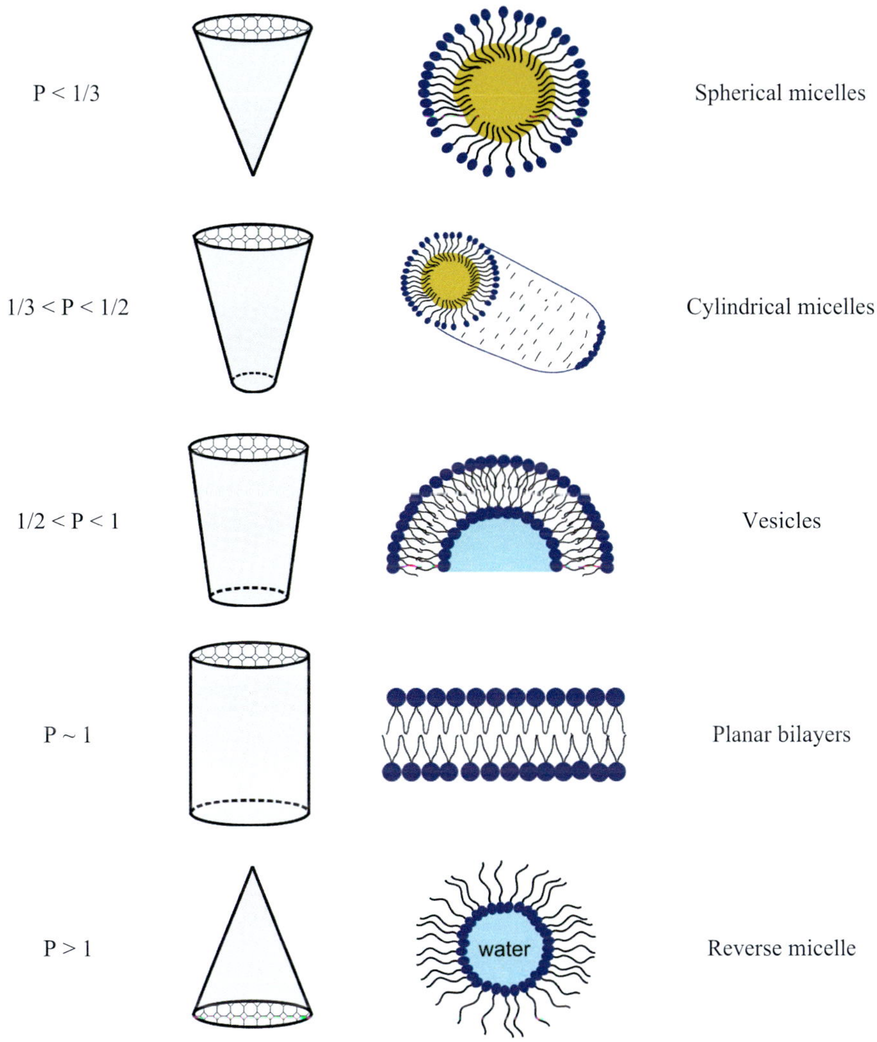

Fig. 8.5 Packing parameter of NDSs in relation to aggregate structures formed

8.5 Utilization of Naturally Derived Surfactants

The surfactant's amphipathic structure and chain length, ionic strength, impurities, approximate molecular weight, pH, temperature, concentration, etc., are the major factors which influence the CMC of a given NDS; hence, NDSs should be purified (to the best level possible) for the purpose of quality control and assured healthy standard and their specification ascertained with requisite condition(s) for replicability before utilization.

Structurally, the hydrophilic portion is made of carbohydrate, glycerol, amino acid, sorbitol, sucrose, phosphate or cyclic peptide, and propylene glycol or polyglycerol; while the hydrophobic portion is composed of long-chain fatty acids or fatty acid derivatives of fats and oils such as palm kernel, rapeseed, soybean, and coconut oils (Ranasalva et al. 2014; Sharma 2014). Most of these NDSs perform additional functions (Table 8.2) in their different applications (WHO 1974; Zobel 1976; Saltmarsh 2020), and some commonly utilized NDSs with their amphipathic structures are presented in Table 8.3. The combination of selected NDSs tends to render more

Table 8.2 Functions of some NDSs in food formulation

S/№	NDS	Functions	References
1	Lecithin	(a) Emulsifier and coating agent in confectionery. (b) Emulsifier, texture improver, anti-spattering agent, and flavor-release agent in fat spreads. (c) Pan-release agent, fermentation stabilizer, and dough conditioner in doughs and bakeries. (d) Dispersant and wetting agent to keep cocoa and cocoa butter in a candy bar from separating. (e) Preservative for baked foods.	Kralova and Sjöblom (2009), Sharma (2014), List (2015), Alhajj et al. (2020), Gombač et al. (2021)
2	Polyoxyethylene sorbitan esters	(a) Pan-release agent for bakery wares. (b) Dough strengtheners in bakery products. (c) Emulsion stabilizer in dairy products, ice creams, whipped cream, and non-dairy cream alternatives. (d) Binder in chewing gum, dietary food supplements supplied in solid, liquid, syrup-type, or chewable forms including capsule and tablet forms. (e) Carrier and solvent for colors, fat-soluble antioxidants, and anti-foaming agents.	Sharma (2014), EFSA (2016), Teixeira et al. (2016), Gombač et al. (2021)
3	Mono- and Diglycerides of fatty acids (MDG)	(a) Dispersants and emulsifiers of coating particles, wetting agents and adhesives over candy surface. (b) Fermentation stability enhancer, loaf volume and texture improver for dough, bread, and bakery goods. (c) Emulsifier and binder in beverages, chewing gum, ice creams, cakes, shortening, whipped toppings, margarine, spreads, peanut butter, confections, jam, jellies, and marmalade. (d) Physical stability enhancer toward creaming in milk beverages. (e) Starch-complexing agents and aerating agents in starch, for example., mashed potatoes, fresh and pre-cooked pasta, and cereals (quick-cook rice).	Kralova and Sjöblom (2009) Sharma (2014), Gombač et al. (2021)
4	Acetic acid esters of MDG	(a) Lubricants and release agents in topping powders, chewing gum base, coatings, cakes, quick-cook rice, bread, biscuits and rusks, processed cereal-based foods, and baby foods for infants and young children.	EFEMA (2019)
5	Lactic acid esters of MDG	(a) Crystal modifiers and emulsifier to stabilize finish coat added to chocolate- and sugar-panned confectionery products. (b) Aeration improver, foam stability, texture and volume enhancer in topping powder, non-dairy cream, chocolate products, processed cereals food, and infant and young children food.	Sharma (2014), EFEMA (2019)
6	Citric acid esters of MDG	(a) Stabilizer and synergists for antioxidants in fat emulsion. (b) Emulsion stabilizer for bakery margarine and shortening in baking. (c) Performance improver for flour, bread dough, or other yeast doughs. (d) Emulsifiers, anti-spattering agents, and stabilizers in margarine, mayonnaise, salad dressings, sauces, etc. (e) Binding enhancer for meat in sausages.	EFEMA (2019)

(continued)

Table 8.2 (continued)

S/№	NDS	Functions	References
7	Mono- and diacetyl tartaric acid esters of MDG	(a) Dough conditioners for all baked products, particularly yeast- leavened products, white bread and rusks, and in ready-mixed flours. (b) Carriers and solvents for colors and food antioxidants. (c) Beverage whiteners. (d) Emulsion stabilizer in dairy products and approved for use in special infant formulae based on crystalline amino acids.	EFEMA (2019)
8	Sucrose esters of fatty acids	(a) Emulsion stabilizer for bread, dressing sauces, mayonnaise-like products, ice creams, infant foods, and special baby formula, including hypoallergenic baby formula products containing hydrolyzed proteins, peptides, or free amino acids. (b) Aerating and starch-complexing agents. (c) Dispersants and stability improver of fat in canned liquid coffee, heat-treated processed meat products, edible ices, chewing gum, sugar confectionery, desserts, and beverage whiteners. (d) Carriers and solvents for colors and fat-soluble antioxidants. (e) Binder in food supplements supplied in solid, liquid, syrup-type, or chewable forms including capsule and tablet forms.	Sharma (2014), EFEMA (2019)
9	Polyglycerol esters of fatty acids	(a) Crystallization inhibitors in the formulation of low-fat margarine, spreads, buttercreams, and breakfast cereals; and to prevent the formation of turbidity of sunflower oils during storage in the oil and fat industry. (b) Aerating agents, crystal modifiers, starch-complexing agents, dough conditioners, humectants, defoaming agents, and anti-spattering agents. (c) Carriers and solvents for colors and fat-soluble antioxidants.	Sharma (2014), EFEMA (2019)
10	Propane-1,2-diol esters of fatty acids	(a) Aerating agent and foam stabilizer in milk and cream analogues, flavored fermented milk products (including heat-treated products), edible ices, sugar confectionery, chewing gum, and beverage whiteners.	EFEMA (2019)
11	Sodium- and calcium stearoyl-2-lactylate	(a) Dough gas retention and stability improver for finer structure in yeast-leavened products. (b) Emulsion stabilizers in breakfast cereals, cookies, crackers, cereal and potato-based snacks, and quick-cook rice. (c) Softener to increase the softness and crumb structure of bread and rolls and maintain softness during storage. (d) Dispersant in coffee creamers to ensure good dispersion.	EFSA (2013), Sharma (2014), EFEMA (2019)
12	Sorbitan fatty acid esters	(a) Emulsifiers, aerating agents, and lubricants in cakes, toppings, cookies, and crackers. (b) Crystallization modifiers for fats.	Sharma (2014) EFEMA (2019)

Table 8.3 Some NDSs with their amphipathic structures

S/N	Name	Structure of surfactant	Surfactant's hydrophobic part with name	Surfactant's hydrophilic part with name
1	Lecithin	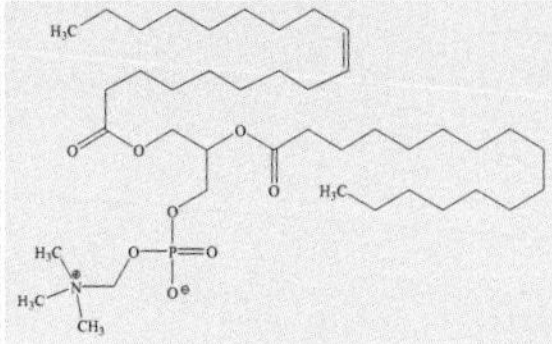	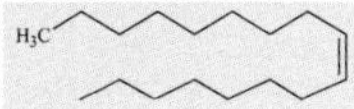 Heptadec-8-ene 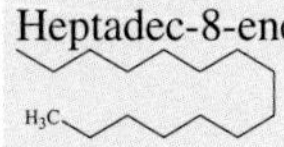 Pentadecane	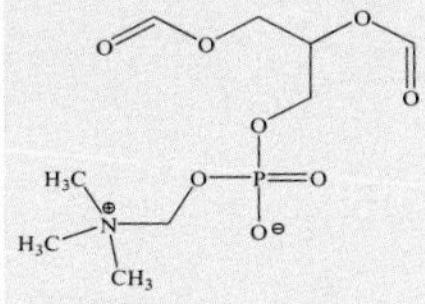 Phosphatidylcholine
2	Polyoxyethylene sorbitan esters	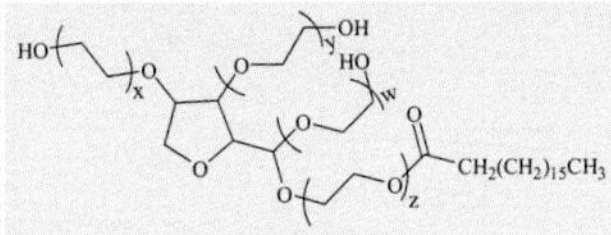	 Heptadecane	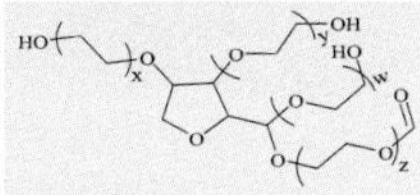 Polyoxyethylene sorbate
3	Monoglycerides of fatty acids and Diglycerides of fatty acids	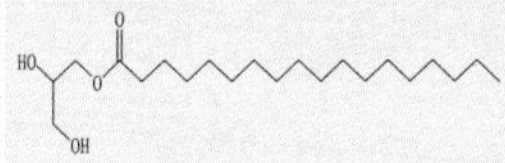 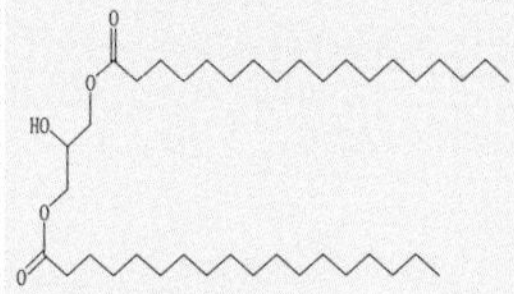	 Heptadecane	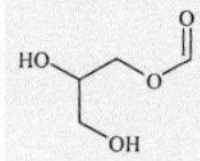 Formic acid 2,3-dihydroxy-propyl ester 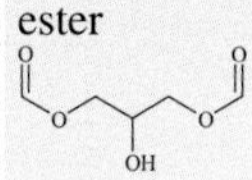 Formic acid 3-formyloxy-2-hydroxy-propyl ester
4	Acetic acid esters of MDG	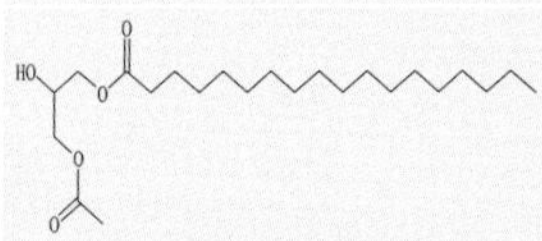	 Heptadecane	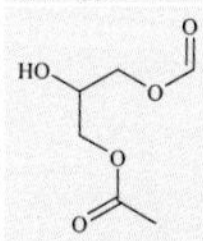 Acetic acid 3-formyloxy-2-hydroxy-propyl ester
5	Lactic acid esters of MDG	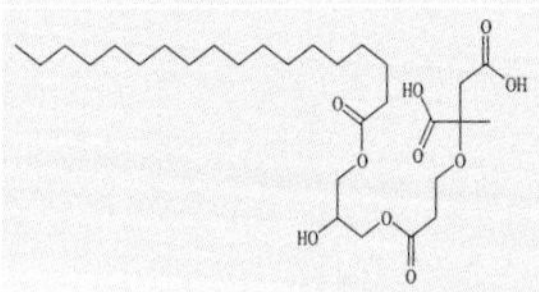	 Heptadecane	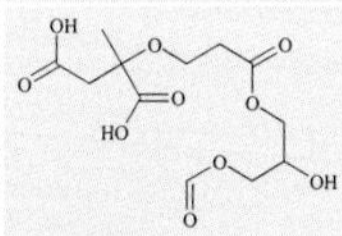 2-[2-(3-Formyloxy-2-hydroxy-propoxycarbonyl)-ethoxy]-2-methyl-succinic acid
6	Citric acid esters of MDG	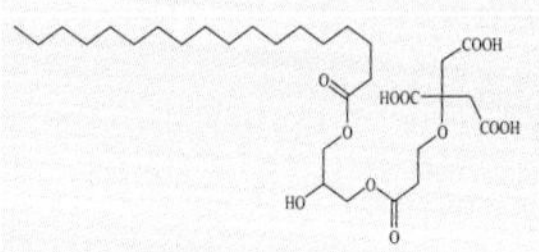	 Heptadecane	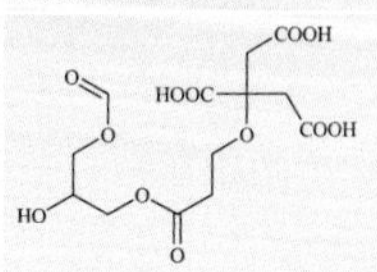 3-Carboxy-3-[2-(3-formyloxy-2-hydroxy-propoxycarbonyl)-ethoxy]-pentanedioic acid

(continued)

Table 8.3 (continued)

S/N	Name	Structure of surfactant	Surfactant's hydrophobic part with name	Surfactant's hydrophilic part with name
7	Mono- and diacetyl tartaric acid esters of MDG	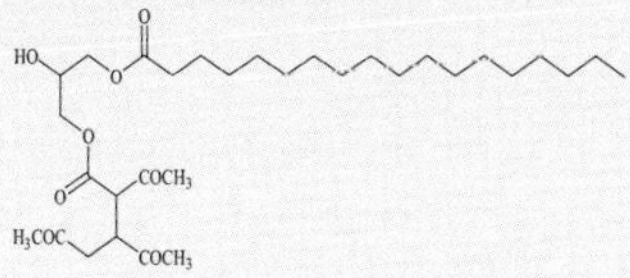	 Heptadecane	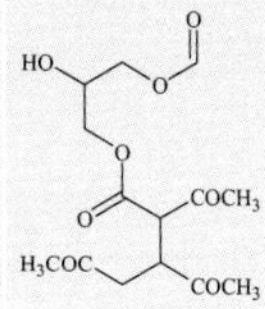 2,3-Diacetyl-5-oxo-hexanoic acid 3-formyloxy-2-hydroxy-propyl ester
8	Sucrose esters of fatty acids	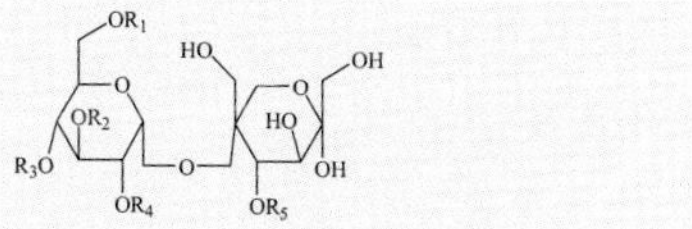 $R_1 = R_2 = R_3 = R_4 = R_5 = CH_2(CH_2)_{15}CH_3$	 Heptadecane	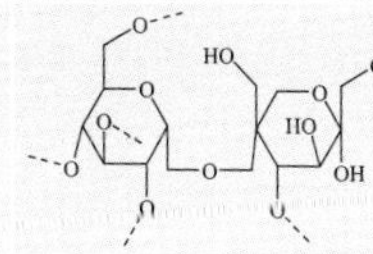 Sucrose unit
9	Polyglycerol esters of fatty acids	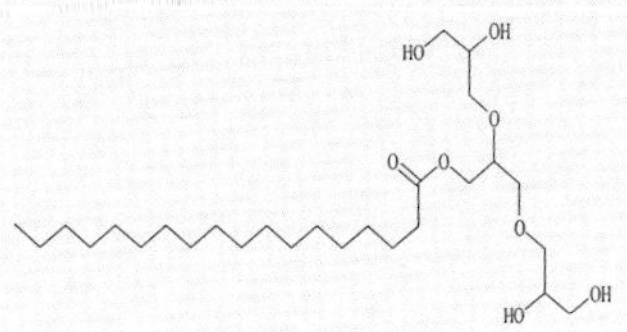	 Heptadecane	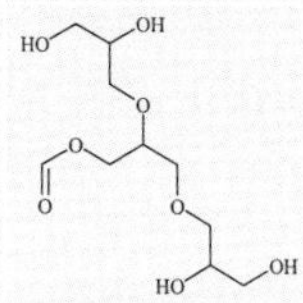 Formic acid 2,3-bis-(2,3-dihydroxy-propoxy)-propyl ester
10	Propane-1,2-diol esters of fatty acids	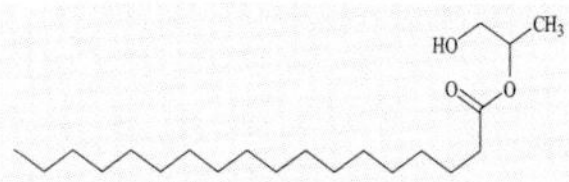	 Heptadecane	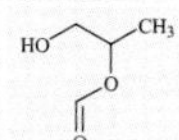 Formic acid 2-hydroxy-1-methyl-ethyl ester
11	Sodium stearoyl-2-lactylate	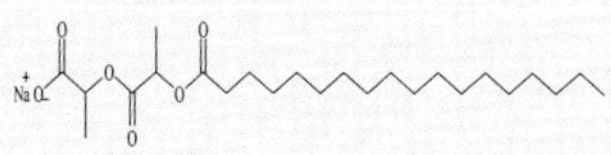	 Hexadecane	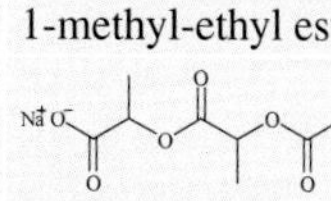 Sodium 2-(2-acetoxy-propionyloxy)-propionate
12	Calcium stearoyl-2-lactylate	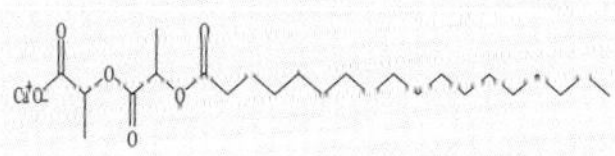	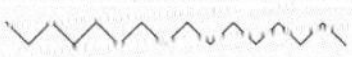 Hexadecane	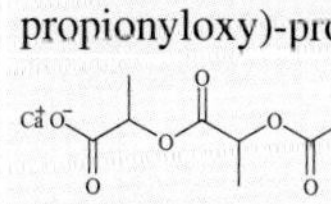 Calcium 2-(2-acetoxy-propionyloxy)-propionate
13	Sorbitan fatty acid esters	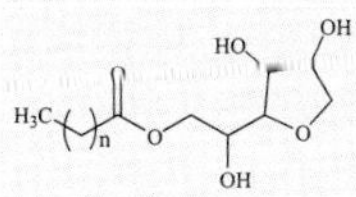	 Polyethylene unit	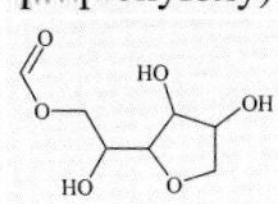 Formic acid 2-(3,4-dihydroxy-tetrahydro-furan-2-yl)-2-hydroxy-ethyl ester

functional amphipathic efficacy as they provide a compatible blend of properties in the formulation of food. For example, a mixture of NDSs composed of mono- and di-glycerides of fatty acids will yield bakery products with stably fermented and well-conditioned dough, which adds nutritive value to the body when consumed (Sharma 2014).

8.6 Conclusion

Naturally derived surfactants patterned for healthy food formulation are a distinct set of organic compounds with peculiar interfacial stabilizing activity between the hydrophilic and hydrophobic portions of the immiscible colloidal system, and as such display their characteristic additive functions in the establishment and development of long-lasting structured edible forms by acting as emulsifiers, dispersants, shelf life-extenders, solubilizers, lubricants, moisture-retainers, quality-improvers, wettability-enhancers, flavor-release agents, crystal-modifiers, foaming agents, and so on. Hence, as human population increases globally with its attendant rise in food requirement, the process of food formulation or creation of special dietary food products with consumable-type NDSs (having structural features identical to those existing in the natural system of the human body) becomes paramount, so as to guarantee sustainable food systems and high level of food security, wellness-quality, availability, digestibility, compatibility, optimal-utility, preservability, nutrimental-activity, sufficiency, and affordability for different classes of the populace (Sustainable Development Goals – 1, 2, 3, and 12).

References

Alhajj MJ, Montero N, Yarce CJ, Salamanca CH. Lecithins from vegetable, land, and marine animal sources and their potential applications for cosmetic, food, and pharmaceutical sectors. Cosmetics. 2020;7(4):87.

Atta DY, Negash BM, Yekeen N, Habte AD. A state-of-the-art review on the application of natural surfactants in enhanced oil recovery. J Mol Liq. 2021;321:114888.

Chhetri AB, Watts KC, Rahman MS, Islam MR. Soapnut extract as a natural surfactant for enhanced oil recovery. Energy Sources Part A Recov Utilization Environ Effects. 2009;31(20):1893–903.

De S, Malik S, Ghosh A, Saha R, Saha B. A review on natural surfactants. RSC Adv. 2015;5:65757–67.

EFSA Panel on Additives and Products or Substances used in Animal Feed (FEEDAP). Safety and efficacy of polyoxyethylene (20) sorbitan monooleate as a feed additive for all animal species. EFSA J. 2016;14(3):4443.

EFSA Panel on Food Additives and Nutrient Sources added to Food (ANS). Scientific opinion on the re-evaluation of sodium stearoyl-2-lactylate (E 481) and calcium stearoyl-2-lactylate (E 482) as food additives. EFSA J. 2013;11(5):3144.

European Food Emulsifier Manufacturers Association (2019). EFEMA index of food emulsifiers.

Gombač Z, Osojnik Črnivec IG, Skrt M, Istenič K, Knez Knafelj A, Pravst I, Poklar Ulrih N. Stabilisation of lutein and lutein esters with polyoxyethylene sorbitan monooleate, medium-chain triglyceride oil and lecithin. Foods. 2021;10(3):500.

Goyal PS, Aswal VK. Micelle structure and inter-micellar interactions in micellar solutions: results of small angle neutron scattering studies. Curr Sci. 2001;80(8):972–9.

Holmberg K. Natural surfactants. Curr Opin Colloid Interface Sci. 2001;6(2):148–59.

Hosseinzadeh R, Khorsandi K, Hemmaty S. Study of the effect of surfactants on extraction and determination of polyphenolic compounds and antioxidant capacity of fruits extracts. PLoS One. 2013;8(3):e57353.

Isaac OT, Pu H, Oni BA, Samson FA. Surfactants employed in conventional and unconventional reservoirs for enhanced oil recovery—a review. Energy Rep. 2022;8:2806–30.

Israelachvili JN, Mitchell DJ, Ninham BW. Theory of self-assembly of lipid bilayers and vesicles. Biochim Biophys Acta. 1977;470:185–201.

Kralova I, Sjöblom J. Surfactants used in food industry: a review. J Dispers Sci Technol. 2009;30(9):1363–83.

List GR. Soybean lecithin: food, industrial uses, and other applications. Polar lipids. 2015:1–33.

Massarweh O, Abushaikha AS. The use of surfactants in enhanced oil recovery: a review of recent advances. Energy Rep. 2020;6:3150–78.

Mehrjoo H, Riazi M, Norouzi-Apourvari S. A comprehensive review on the use of eco-friendly surfactants in oil industry. Chemical Methods. 2022:357–99.

Morton S, Pencheon D, Neil Squires N. Sustainable development goals (SDGs), and their implementation: a national global framework for health, development and equity needs a systems approach at every level. Br Med Bull. 2017;124:81–90.

Nagarajan R. Molecular packing parameter and surfactant self-assembly: the neglected role of the surfactant tail. Langmuir. 2002;18(1):31–8.

Nwaichi EO, Ntorgbo SA. Assessment of PAHs levels in some fish and seafood from different coastal waters in The Niger Delta. Toxicol Rep. 2016;3:167–72.

Ranasalva N, Sunil R, Poovarasan G. Importance of biosurfactant in food industry. IOSR J Agric Veterinary Sci. 2014;7(5):06–9.

Saltmarsh M. Food additive regulations in Europe. London: Saltmarsh's Essential Guide to Food Additives; Royal Society of Chemistry; 2020. p. 40–51.

Schramm LL, Stasiuk EN, Marangoni DG. Surfactants and their application. Annual Report on the Progress of Chemistry, Section C. 2003;99:03–48.

Sharma RK. Surfactants: basics and versatility in food industries. PharmaTutor. 2014;2(3):17–29.

Stuart MCA, Boekema EJ. Two distinct mechanisms of vesicle-to-micelle and micelle-to- vesicle transitions are mediated by the packing parameter of phospholipid-detergent systems. Biochim Biophys Acta. 2007;1768:2681–9.

Teixeira GL, Züge LCB, Silveira JLM, Scheer ADP, Ribani RH. The impact of polyoxyethylene sorbitan surfactants in the microstructure and rheological behaviour of emulsions made with melted fat from Cupuassu (Theobroma grandiflorum). J Surfactant Deterg. 2016;19(4):725–38.

World Health Organization (1974). Toxicological evaluation of some food additives including anticaking agents, antimicrobials, antioxidants, emulsifiers and thickening agents.

Yan Y, Xiong W, Xiaosong L, Lu T, Huang J, Li Z, Fu H. Molecular packing parameter in bolaamphiphile solutions: adjustment of aggregate morphology by modifying the solution conditions. J Phys Chem B. 2007;111:2225–30.

Zobel M. Toxicological evaluation of some food additives including anticaking agents, antimicrobials, antioxidants, emulsifiers and thickening agents. WHO food additives series, no. 5. 520 seiten Geneva 1974. Preis: Sw. Fr. 23; 1976: pp. 681–682.

Rethinking Agenda 2063: Leveraging STEM Women's Empowerment for Food Security in a Post-COVID-19 Pandemic Era

9

A. Henri-Ukoha and I. I. Ukoha

9.1 Introduction

During the golden jubilee celebration of the formation of the Organization for African Unity/ African Union, African leaders made a declaration of their commitment to support Africa's new path for attaining inclusive and sustainable economic growth and development. This led to the development of Agenda 2063 which is a blueprint and master plan for the transformation of Africa into a global driving force of the future. Agenda 2063 is Africa's strategic framework that aims to deliver on its goal for inclusive and sustainable development. Agenda 2063 seeks to deliver on a set of *seven aspirations*, each with its own set of goals which if achieved will move Africa closer to achieving its vision for the year 2063. These seven aspirations reflect our concrete manifestation of the pan-African drive for shared prosperity and well-being; unity and integration; and freedom and expanded horizons for the citizens, where the full potential of women and youth are realized, with freedom from fear, disease, and want (African Union 2021). Africa plans to achieve this vision within a 50-year period from 2013 to 2063 through Agenda 2063.

Aspiration 1 of Agenda 2063 foresees "a prosperous Africa based on inclusive growth and sustainable development." To achieve this ambition, one of the key goals (goal 5) for Africa (Nigeria inclusive) is to ensure the transformation of agriculture for increased production and to ensure that its citizens are food secure. Food security means having, at all times, both physical and economic access to sufficient, safe, and nutritious food to meet dietary needs and food preferences for a productive and healthy life (Food and Agriculture Organization, United States AID, USAID 2021). Food security goes far beyond having enough food. It is being able to access food of high nutritional quality regularly, without worrying about not having food on the table. There are four components of food security. These include *availability, affordability/access, utilization, and stability.*

Availability: This is having sufficient and consistent quantities of appropriate food available. Such food can be supplied through household production, other domestic outputs, commercial imports, or food assistance.

Affordability/access: This is about having adequate income or other resources to access appropriate food for a nutritional diet. Access depends upon the income available to the household, on the distribution of income within the household, and on the price of food.

A. Henri-Ukoha (✉)
Department of Agricultural Economics and Agribusiness Management, University of Port Harcourt, Choba, Rivers State, Nigeria
e-mail: adanna.henri-ukoha@uniport.edu.ng

I. I. Ukoha
Department of Agricultural Economics, Federal University of Technology, Owerri, Imo State, Nigeria

E. O. Nwaichi (ed.), *Modernity in Health and Disease Diagnosis: The Account from STEM Women*, Sustainable Development Goals Series, https://doi.org/10.1007/978-3-031-34963-8_9

Utilization/consumption: This is having adequate dietary intake and the ability to absorb and use nutrients in the body (FANTA n.d.). Effective food utilization is mostly dependent on knowledge within the household of food storage and processing techniques, basic principles of nutrition, and proper childcare (FAO 2011).

Stability: This is the ability to have access to adequate food at all times.

A family is food secure when its members do not live in hunger or fear of hunger. Food and nutrition security issues occupy a central place on the global agenda for sustainable development and Agenda 2063.

It is worrisome to realize that between 720 and 811 million people, the majority of whom live in Africa, still go to bed hungry each night, in 2020 in particular (FAO 2021). The World Food Programme estimates that this number could exceed 1 billion (WFP 2021). Africa is the region with the highest prevalence of undernourishment in percentage terms, at 21 percent (WFP 2021). About 257 million people with an exponential rise of 32.6 million are malnourished in sub-Saharan Africa, and more than half of this figure is from West Africa (FAO and Economic Commission of Africa, ECA 2018).

This indicates that the food security situation in West Africa is even more alarming as between 36 and 40 million people in this region are estimated to be food insecure. The prevalence of undernourishment rose from 10.4% in 2010 to 14.7 percent in 2018 (FAO 2019). In 2019, over 821 million people were food insecure, and the prevalence of undernourishment rose from 8.4 percent in 2019 to 9.9%. This report indicates a worsening food insecurity scenario. The total population of the West African sub-region is projected to reach 420 million in 2020 (Zoungrana 2013). In order to feed a population expected to grow to 9 billion people by 2050, the world will have to double its current food production (United States AID, USAID 2021). The number of food insecure people is expected to reach 12.9 million. This scenario was primarily driven by unsustainable food systems, conflict and social unrest, socioeconomic conditions, natural hazards, climate change, and pests and is now compounded by the COVID-19 pandemic.

SARS-CoV-2 was first reported in Wuhan city of China in December 2019 (World Health Organization 2020). The disease is caused by a zoonotic respiratory epidemic that has been declared by the World Health Organization (WHO) as a global public health emergency. COVID-19 disease and the fear of disease generated extensive global economic and social impacts, along with restrictions on international travel imposed by most countries, the quarantining of millions of people, and disruption of supply chains for food and manufactured products (Nicola et al. 2020). The agricultural sector is not left behind as COVID-19 disrupted many activities in fisheries, livestock, crops, and their supply chains (McNamara et al. 2020; Gortazar and de la Fuente 2020). In fact, COVID-19 affects agriculture in both the supply and demand for food (FAO 2020), placing food security at risk in many key aspects of the food system value chain. Inflation coupled with job losses (OCHA. Global Humanitarian Response Plan Covid-19 2020) due to COVID-19 and persistent income inequality continue to exacerbate food insecurity and malnutrition (WFP 2021).

Households are also affected as the COVID-19 pandemic is creating worrisome impacts on their incomes and food supply chains. For instance, movement restrictions have reduced the availability of migrant labor, interrupting some harvesting and agricultural activities, increasing levels of post-harvest losses due to a reduced workforce, and delaying the delivery of fresh produce to various target markets (OECD 2020).

The use of quarantines, bans, and restrictions on the movement of goods and people as disease control measures has resulted in significant socioeconomic consequences for livelihoods, especially for poor rural farmers in developing countries (Nicola et al. 2020). The COVID-19 pandemic is estimated to push half a million communities in developing countries into poverty (Sumner et al. 2020). Poor countries are the worst-hit by the pandemic as additional 54 million and 24 million people, respectively, will live below the poverty line and go into extreme poverty (Azcona et al. 2020).

The story is the same in Nigeria as World Food Programme reported a 34.1 percent loss in its gross domestic product due to COVID-19, which amounts to USD 16 billion. The agriculture sector in particular, which serves as the primary means of livelihood for most Nigerians, suffered a 13.1% loss in output amounting to USD 1.2 billion during the lockdown periods. Unfortunately, during the lockdown periods alone, 27 million additional people fell below the poverty line in Nigeria. Broadly speaking, 40% of Nigerians (83 million people) live below the poverty line, while another 25% (53 million) are vulnerable due to the pandemic. With COVID-19, many of these 53 million vulnerable people could fall into poverty. A devastating reality is that the path to zero hunger through goal I of Agenda 2063 is being stopped dead in its tracks by COVID-19.

World Food Programme estimates that *272 million people are already at risk of becoming acutely food insecure due to the aggravating effect of the COVID-19 crisis* (WFP 2021). This scenario poses a serious threat to Agenda 2063. With the COVID-19 pandemic ravaging the whole world, food insecurity will likely remain for some time in the near future (FAO 2011). The report by OECD corroborates that the pandemic has had and will continue to have a major impact on the health and well-being of many vulnerable groups (OECD 2020). Therefore, our collective ability to meet zero hunger by 2063 appears elusive unless urgent actions are taken. The ultimate solution to combating hunger and food insecurity at the national as well as the global level is to empower people with opportunities to earn adequate income and to assure an abundant supply of food from either domestic production or imports, or both.

Women are more adversely affected by the social and economic impacts of the COVID-19 pandemic, including losing livelihoods and experiencing decreases in their personal incomes. This is because women are more likely to experience food insecurity than men in more countries worldwide (UN Women 2018). Unfortunately, women constitute 43 percent of the agricultural labor force in developing countries and, despite their contribution to half of the world's food, often face constraints that limit profits, contributions, and productivity (UNDRR 2019). Increasing women's access to resources and opportunities could substantially reduce the number of hungry people in the world. No wonder, FAO reported that increasing women's contribution to food production and enterprise could reduce the number of hungry people in the world by 12–17%, or by 100 to 150 million people (FAO 2011). Hence, empowering women and the food systems that nourish them in this era is more important than ever.

"Women's empowerment means enhancing women's sense of worth, their right to have and to determine choices, their right to have access to opportunities and resources, their right to have the power to control their own lives both within and outside the own and their ability to influence the direction of the social change to create a more just economic order nationally and internationally" (United Nations Population Information Network 2n.d). The International Food Policy Research Institute (IFPRI) created an innovative measure called the Women's Empowerment in Agriculture Index (WEAI) which is used to capture women's empowerment. According to them, it is categorized into five dimensions which include: decisions about agricultural production, access and decisions over productive resources, control over the use of income, leadership, and time use (Salazar and Fahsbender 2019).

"More involvement of women in the creation of post-COVID future is critical in building back better food systems where there is equal access to nutritious food and decent livelihoods" (International Fund for Agricultural Development), a vision for sustainable development goal 3.

"Therefore, when women and girls have better access to information, resources and economic opportunities, and are free to make their own decisions, hunger rates fall and nutrition improves not only for themselves but also their families, communities and countries". Empowering them will contribute to the recovery from the

COVID-19 pandemic and in creating an environment that will reduce poverty, enhance productivity, and improve food security.

9.2 How Can this be Achieved?

Agenda 2063 can be achieved in a post-COVID era through the empowerment of women, which will be considered in two dimensions: direct empowerment of STEM women and the transfer of knowledge to others who will benefit from the empowered STEM women.

9.2.1 Direct Empowerment of STEM Women

The women can be empowered in the following areas:

1. Empowerment in Knowledge Sharing and Participation.
   Food security can be achieved through the involvement and active participation of women in educational, capacity building, and knowledge transfer activities and projects on technologies in agriculture and sustainable practices. Although the role of women in agriculture cannot be over-emphasized, there are still a significant number of initiatives that are directed at men. Incentivizing the adoption of advanced technologies among STEM women will also boost agricultural production.
2. Empowerment Through the Formation of STEM Women Agripreneurs.
   STEM women can establish the Women Agripreneurs-in-STEM (WASTEM) program. STEM women who want to start their own innovative business build and grow your dream business especially in agripreneurship can be trained, empowered, and supported through this platform. Empowering women through participatory approaches will allow them to identify specific actions to promote and recognize their roles in agriculture toward increased productivity vis-à-vis food security. Studies have shown that for every dollar raised, women generate 78 percent of revenue compared to 31 percent from their male-run start-ups. This will undoubtedly enhance their income level, thereby increasing their level of food security in the post-COVID-19 pandemic era. Innovative entrepreneurship contributes to the wealth of nations and their economic dynamism through, for instance, job creation.

   Training and empowerment can be in the areas of aquaponics (using fish/aquatic waste to grow plants, through a cyclical system), hydroponics [the process of growing plants in a medium (without soil) and adding nutrients], aeroponics (the process of *growing* plants in an air or mist environment without the use of soil or an aggregate medium), bucket farming, bag farming, portable snail, goat, fish, and battery cage system among others. It can also be in the area of exposing them to the opportunities of participating as actors in the agricultural value chain for agricultural development and food security. They could get involved in crop production, livestock production, agroforestry, and fisheries, among others. They can also get involved in the processing and marketing/distribution of any of these enterprises in addition to their jobs. In Australia, STEM women are positioning themselves as key players in agriculture by applying their skills in STEM (Science, Technology, Engineering, and Mathematics) to give them that competitive edge to improve health and well-being (SDG3) through quality food.

   Mentorship can also be provided to the agripreneurs, through which STEM women can access a whole stream of insightful online events focused to address key issues affecting the sector.
3. Empowerment with Financial Resources.
   This can be achieved through the development of partnerships with government, financial institutions, non-governmental agencies, and other relevant stakeholders. It will also involve the design and development of programs that facilitate STEM women agripreneurs' access to financial resources,

especially from financial institutions and other concerned agencies.

4. Empowerment to Establish Group Farms.
UniPort STEM women can be encouraged to establish group farms within or outside the university environment. This will lead to increased agricultural productivity. The income generated from the sales of the proceeds when sold to the members of the university community obviously translate to food security towards the attainment of Agenda 2063 in this post-COVID-19 pandemic era.

9.2.2 Empowered STEM Women Transfer the Knowledge and Skill to Others

This can be achieved in the following areas:

1. Formation of Young Farmers' Club for STEM Girls.
The formation of young farmers' clubs in secondary schools particularly for STEM girls will stimulate their interest in agriculture. This will help to "catch them young." Empowerment can be through the exposure of students to innovative agricultural practices and agriculture potential information thereby raising a generation of students who are willing to get involved in agriculture. This will help to improve agricultural production. Moreso, the empowered STEM women train the students on how to manage farms and regularly organize agricultural science debates and competitions for them. This is geared toward the achievement of food security in a post-COVID-19 pandemic era.
2. Empower Women in University Host Communities.
Women in the communities are key agents for achieving the transformational economic, environmental, and social changes required for sustainable development. The STEM women can reach out to the women farmers in the university host communities and empower them through capacity building on new, innovative, better, and sustainable farming practices. Training equips them with skills to pursue new livelihoods, increase their productive potential, and adapt technology to their needs. Empowering them is key not only to their well-being, households, and communities but also to overall economic productivity. Evidence shows that this spurs productivity gains, enhanced growth, and improved development prospects for current and future generations. Empowering women economically and socially can enhance economic growth and food security and sustainable development. "If women had the same access to productive activities as men, agricultural production would increase, resulting in the feeding of approximately 150 million more people" (FAO 2011). The empowered STEM women can partner with government and non-governmental agencies and other concerned stakeholders for the provision of such farm inputs as fertilizers, pesticides, high-yielding crops, and livestock varieties at subsidized rates. Training can also be a better way of marketing their farm produce as well as for easy and affordable access to loans and credit facilities.
   Women in university host communities can also be exposed to better farming methods through exposure visits, agricultural shows, and exhibitions. This will help to improve productivity and nutritional quality.
   Women can be trained in group dynamics and encouraged to form small groups for easy training, mentoring, financing, and monitoring. This will promote their income-earning opportunities along the value chain, toward the attainment of food security.
3. Organizing Career Guidance and Counseling in Secondary Schools.
The empowered STEM women, through the organization of career guidance and selling in secondary schools, will share insights, lessons, and tips to motivate and inspire students to make informed decisions about their future. The career talks will allow students to hear from real-life role models who can demonstrate the relevance and connection of what is taught in the classroom. It will also equip stu-

dents with the inspiration and information to make better-informed decisions about their future career paths. This will undoubtedly promote food security.

4. Link and Learn Online Social Media Platforms on Agriculture.

   The empowered STEM women will establish links and learn online social media platforms. This will be achieved through organizing knowledge sharing and networking events, training, and webinars for the agripreneurs. This platform will connect agripreneurs from all walks and provide them the opportunity to widen their network and interact with peers from other environments. This will help to connect with new talent from a wide variety of backgrounds, showing them that agriculture is a field in which they can make a great impact on food security.

Through this platform, mentorship can be provided that can challenge and provide mentorship to upcoming agripreneurs to help them grow their agribusiness.

9.3 Conclusion

Achieving zero hunger in the year 2063 in the post-COVID-19 pandemic era will require new and existing applications in science, technology, and engineering across the food system, addressing all dimensions of food security. The empowerment of STEM women is critical, not only for ensuring food at all times but also for harnessing agriculture and the broader food system as a driver of Agenda 2063.

References

African Union. Our Aspirations for the Africa We Want. Agenda 2063. 2021. https://au.int/en/agenda2063/aspirations

Azcona G, Bhatt A, Encarnacion J, Plazaola-Castano J, Seck P, Staab S, Turquet L. From Insights to Action: Gender Equality in the Wake of COVID-19. New York, NY: UN Women; 2020.

FAO. COVID-19 and the Risk to Food Supply Chains: How to Respond? Rome: FAO; 2020. p. 7.

FAO. Empowering women and girls is crucial to ensure sustainable food security in the aftermath of COVID-19, say UN food agency heads on International Women's Day. 2021.

Food and Agriculture Organization. Women and food security; 2011.

Gortazar C, de la Fuente J. COVID-19 is likely to impact animal health. Prev Vet Med. 2020;180:105030. https://doi.org/10.1016/j.prevetmed.2020.105030.

McNamara T, Richt JA, Glickman L. A critical needs assessment for research in companion animals and livestock following the pandemic of COVID-19 in humans. Vector Borne Zoonotic Dis. 2020;20:393–405. https://doi.org/10.1089/vbz.2020.2650.

Nicola M, Alsafi Z, Sohrabi C, Kerwan A, Al-Jabir A, Iosifidis C, et al. The socio-economic implications of the coronavirus pandemic (COVID-19): a review. Int J Surg. 2020;78:185–93. https://doi.org/10.1016/j.ijsu.2020.04.018.

OCHA. Global Humanitarian Response Plan Covid-19. Geneva: United Nations Office for the Coordination of Humanitarian Affairs (OCHA); 2020.

OECD. COVID-19 Policy Brief on Well-being and Inclusiveness. 2020. http://www.oecd.org/coronavirus/en/

Salazar L, Fahsbender J. Improving food security through women's empowerment. The Agrifood Support Program (APAGRO); 2019. https://blogs.iadb.org/sostenibilidad/en/improving-food-security-through-womens-empowerment/

Sumner A, Hoy C, Ortiz-Juarez E. Estimates of the impact of COVID-19 on global poverty. In: WIDER Working Paper 2020/43. Helsinki: UNU-WIDER; 2020. Available online at: https://www.wider.unu.edu/sites/default/files/Publications/Working-paper/PDF/wp2020-43.pdf.

UN Women. UN Women Annual Report 2017–2018 | UN Women. 2018. https://www.unwomen.org.digital-library.publications

UNDRR. Gender research in IPM: Women's empowerment as a key to unlocking food security. 2019.

United States AID, USAID. Agriculture and Food security. 2021.

WFP. World Food Programme: Nigeria. Country Brief April 2021. 2021.

World Health Organization. Intensive prevention and control during health care when Covid-19 is suspected. 2020.

Antimicrobial Resistance: A Collective Responsibility

10

Kome Otokunefor, Obakpororo Ejiro Agbagwa, Esther Oshifade, Chidozie Ngozi Endarlyn Ibezim, and Esther Azi

10.1 Introduction

The World Health Organization (WHO) has declared the scourge of Antimicrobial Resistance (AMR) a global emergency of gigantic proportions (WHO 2018). Antimicrobial resistance refers to the ability of a pathogen to withstand the effects of an antimicrobial agent, leading to reduced drug effectiveness. Historically, for every class of drugs developed, bacterial resistance was observed within a very short time frame (Fig. 10.1), with this occurring within two years of the introduction of some of these drugs into clinical practice (Peterson and Kaur 2018). Over the years, drug-resistant pathogens have given rise to multidrug-resistant pathogens exhibiting resistance to three or more antimicrobial drug classes. The final evolution in drug resistance appears to be the pan-resistant bacteria which are resistant to drugs in every classification. This scenario is reminiscent of a return to the pre-antibiotic era, where mortality/death rates were up to 80% (Dhingra et al. 2020).

Currently, estimates reckon that antimicrobial resistance is responsible for up to 700,000 deaths worldwide, with one person dying from an antibiotic-resistant infection every 45 seconds (Wall 2019). It is presently projected that by the year 2050, AMR will be responsible for a staggering ten million deaths globally each year (O'Neill 2016), with a global loss of $100 trillion, and that Africa (along with India) will be one of the two worst places to be hit, with an estimated 4.1 million deaths per year. To put this in perspective, by 2050, antimicrobial resistance is expected to be the leading cause of death globally – far exceeding cancer slated to be the second most common cause responsible for 8.2 million deaths.

K. Otokunefor (✉) · O. E. Agbagwa
Department of Microbiology, Faculty of Science, University of Port Harcourt, Port Harcourt, Nigeria
e-mail: kome.otokunefor@uniport.edu.ng; obakpororo.agbagwa@uniport.edu.ng

E. Oshifade
Department of Animal Science, Faculty of Agriculture, University of Port Harcourt, Port Harcourt, Nigeria

C. N. E. Ibezim
Department of Pharmaceutical Microbiology and Biotechnology, Faculty of Pharmaceutical Sciences, University of Port Harcourt, Port Harcourt, Nigeria
e-mail: chidozie.ibezim@uniport.edu.ng

E. Azi
Department of Community Medicine, University of Port Harcourt Teaching Hospital, Port Harcourt, Nigeria

E. O. Nwaichi (ed.), *Modernity in Health and Disease Diagnosis: The Account from STEM Women*, Sustainable Development Goals Series, https://doi.org/10.1007/978-3-031-34963-8_10

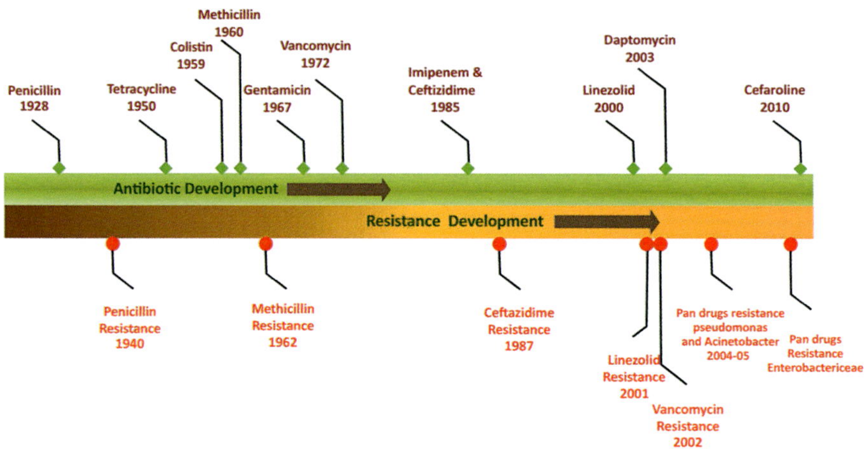

Fig. 10.1 Overview of the rate of resistance development in comparison to the development of key antibiotics. (Source: https://www.intechopen.com/chapters/65640)

10.2 Antimicrobial Resistance Development

The phenomenon of antimicrobial resistance is multidisciplinary, arising from several possible driving factors which can generally be classed as misuse, abuse, and overuse. Traditionally, hospital environments have been the major driving factor for the development of antimicrobial resistance (Hernando-Amado et al. 2020) with up to 75% of drug-resistant bacteria found to be associated with healthcare-associated environments (Cassini et al. 2019). The development of this antimicrobial resistance has been found to be linked with an increase in the consumption of antimicrobial agents taking place worldwide. Other factors driving this development include the irrational use of antimicrobials in agriculture (Fig. 10.2). This is a key contributor to the development of antimicrobial resistance. Agricultural antibiotics specifically comprise the bulk of antibiotics consumed in the United States. These agricultural antibiotics are used both as growth promoters found in sub-inhibitory concentrations in animal food, as prophylaxis and in animal therapy (Dadgostar 2019), as well as environmental pollution by antibiotics via agricultural practices, treatment facilities, and pharmaceutical organizations (Qu et al. 2019).

All of these either select for antimicrobial-resistant pathogens or create an enabling environment for the dissemination of antimicrobial-resistant genes to susceptible populations. There are several key mechanisms by which microorganisms develop resistance to an antimicrobial agent. These methods indicated in Fig. 10.3 are varied and often highlight just how "smart" microorganisms are in proving good health and well-being in line with SDG 3.

The overuse, abuse, and misuse of antibiotics are often thought to be some of the driving factors causing the development of resistance. And one of the ways this happens is by the selection of resistant organisms (Fig. 10.4). Following the inappropriate uptake of an antibiotic, there is a greater chance that resistant organisms would be selected. In the absence of other non-resistant organisms, these resistant organisms then have a chance to grow and multiply at a higher frequency than if the selective pressure of the non-resistant organisms were present, thereby posing more of a problem in subsequent disease conditions. Additionally, the resistant organisms have the ability to transmit the factors responsible for this resistance to other non-resistant organisms, recruiting them to the resistant category.

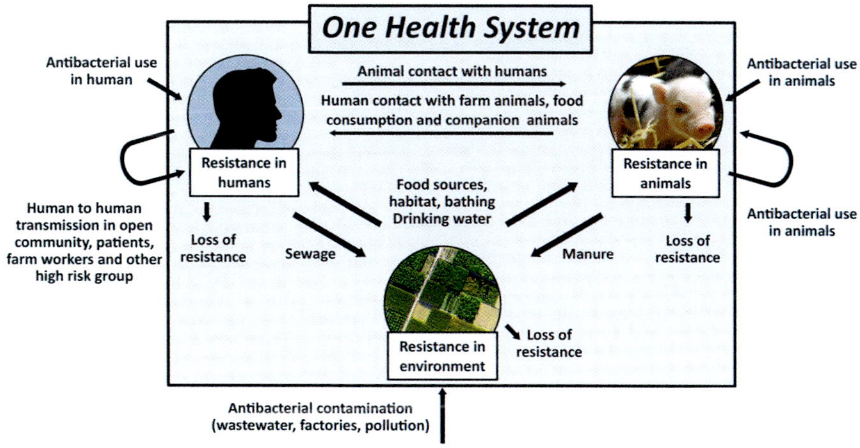

Fig. 10.2 Role of different sectors in the development of antibiotic resistance (Source: Booton et al. 2021)

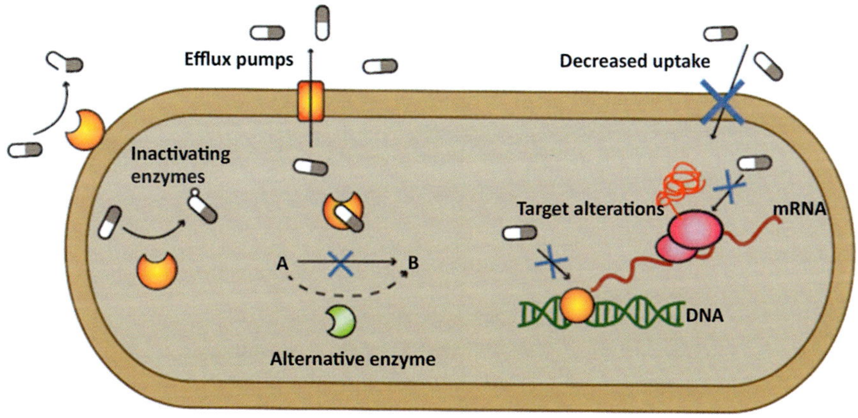

Fig. 10.3 Mechanisms of antibiotic resistance development (Source: Mutuku et al. 2022)

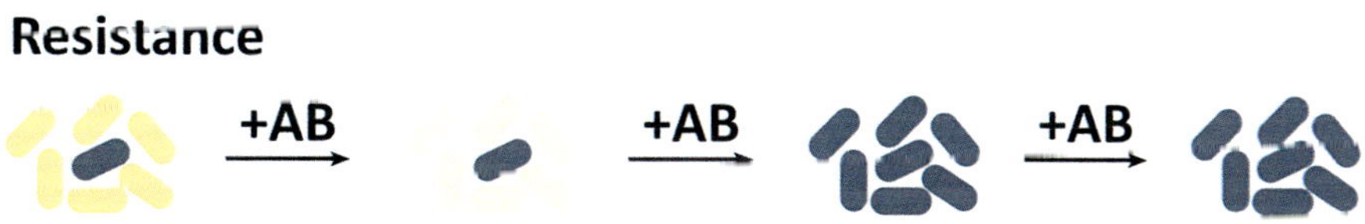

Fig. 10.4 A graphical representation of the process for selection of antibiotic-resistant organisms (Source: Dewachter et al. 2019)

10.3 Consequences of Antimicrobial Resistance

The scourge of antimicrobial resistance is costly. The pathway to antibiotic development has been greatly hampered by the emergence of resistant strains such as methicillin-resistant *Staphylococcus aureus* (MRSA) and resistant *Pseudomonas aeruginosa* with their accompanied treatment challenges (Davies and Davies 2010). This phenomenon has been associated with increased hospital stays and treatment costs, increased mortality and global costs, decreased productivity, and reduced quality of life (Founou et al. 2017). As far back as 2010, a study carried out in the United States, exploring variations in healthcare costs between methicillin-resistant and methicillin-susceptible strains of *Staphylococcus aureus* (MRSA and MSSA, respectively) (Filice et al. 2010), reported a more than 100% increase in the cost of $35,000 used to treat MRSA infections as opposed to the $16,000 used to treat their susceptible MSSA counterparts. Consequently, higher mortality rates of 24% were also associated with MRSA unlike the 11.5% associated with MSSA. This is simply an illustration of how antimicrobial resistance impacts the individual and society at large. In addition to resistance associated with *Escherichia coli*, most problematic nosocomial cases of bacterial antimicrobial resistance are limited to a few bacterial families generally described as ESKAPE pathogens. This acronym consists of *Enterococcus faecium, Staphylococcus aureus, Klebsiella pneumoniae, Acinetobacter baumannii, Pseudomonas aeruginosa*, and *Enterobacter* spp. These organisms specifically are associated with higher levels of mortality, hospital stay length, and cost (Mulani et al. 2019). Several strains of these bacteria are currently termed "superbugs." These groups of organisms are generally resistant to the last-line drugs available to combat them, making them essentially untreatable. Examples of superbug variants include methicillin-resistant *S. aureus*, carbapenem-resistant *Acinetobacter baumannii* and *Pseudomonas aeruginosa*, vancomycin-resistant *Enterococci*, and colistin-resistant *Escherichia coli*.

10.4 Antimicrobial Susceptibility: The African Story

An assessment of the current situation of antimicrobial susceptibility in the African continent with a focus on *E. coli* and the ESKAPE pathogens shows high levels of resistance. A 2017 review article on antimicrobial resistance in West Africa, of which over 50% of articles were from Nigeria, reported pooled resistance rates of up to 81% observed in *Escherichia coli* against some antibiotics (Bernabe et al. 2017). To put this in perspective, out of ten people with *E. coli* infections, those specific drugs would be ineffective in about eight individuals. Similar high levels of resistance were observed for *Klebsiella pneumoniae* and *Staphylococcus aureus* against some antibiotics. On a positive note, however, for the newer third-generation cephalosporins, a higher level of activity was observed against the organisms (i.e., less resistance) with pooled resistance rates as low as 11.9% reported.

A 2018 *Nature Reviews* on MRSA (Lee et al. 2018) reports resistance rates ranging from 25% to 49% from Nigeria. Another 2018 study (Abubakar and Sulaiman 2018) reported rising rates of MRSA in Nigeria between 2009 and 2013 (18.3–42.3%). Once again, to put this in perspective, nearly five out of ten individuals with a *S. aureus* infection would probably encounter treatment failure. Data for vancomycin-resistant *Enterococci* in Nigeria noted resistance rates of up to 88.9%, though on average the pooled rate was 25.3% (Wada et al. 2020).

10.5 Combating Antimicrobial Resistance

With all these staggering statistics and the potential for even ever-rising levels of resistance, control strategies to combat both the development and spread of antimicrobial resistance are essential. These strategies are often aimed to detect and prevent the development of resistance to achieve. As the multifactorial nature of antimicrobial resistance is now accepted, a One Health approach needs to be adopted to combat it. This approach simultaneously tackles the development and spread from three focal points: humans, animals, and the environment (Hernando-Amado et al. 2020). A collaborative approach at the national and international levels between governmental and non-governmental agencies, committees, professional bodies, industries, research institutes, and other relevant stakeholders is paramount in achieving an AMR-free population

(Rajesh et al. 2014). A 2016 review on antimicrobial resistance (O'Neill 2016) commissioned by the then UK prime minister lists out nine interventions needed to combat antimicrobial resistance. The first of these is focused on sensitization involving educating the public on the growing menace of antimicrobial resistance. Also, infection prevention & control (IPC) is notably a natural and assured strategy to curtail the emergence and spread of antibiotic resistance. However, this strategy can only be achieved through immunization and vaccination, adequate environmental hygiene/sanitation, and surveillance. According to CDC, identification of antibiotic resistance infections following antimicrobial resistance studies, etiology of the infection, and risk factors can contribute to the development of strategies to combat such infection and resistance development arising from its treatment. Other proposed intervention strategies involve the improvement in sanitary conditions to prevent/reduce the spread of infectious disease, reducing misuse, abuse, and overuse in humans, animals, and agriculture. Additionally, surveillance and strategies promoting the development of rapid diagnostic techniques, vaccines, and novel antimicrobial agents are key interventions. Antimicrobial stewardship (AMS) is a key part of this process. AMS can be linked to a set of defined, multifaceted, structured, and integrated measures that ensures adequate and appropriate safe use of antimicrobials in order to improve clinical outcomes and minimize further development of AMR.

Antimicrobial stewardship programs have a number of roles to play, some of which include:

- Optimization of the use of antibiotics.
- Promote behavior change in antibiotic prescribing and dispensing practices.
- Improve the quality of care and patient outcomes.
- Save on unnecessary healthcare costs.
- Reduce further emergence, selection, and spread of AMR.
- Prolong the lifespan of existing antibiotics.
- Limit the adverse economic impact of AMR.
- Build the best-practices capacity of healthcare professionals regarding the rational use of antibiotics (Davey et al. 2013; Schuts et al. 2016).

No doubt, combating antimicrobial resistance involves multidisciplinary, innovative, collaborative as well as regulatory mechanisms which need every part of it functioning effectively to be successful. Premanandh and colleagues in a Premanandh et al. 2016 publication succinctly capture this in one diagram (Fig. 10.5).

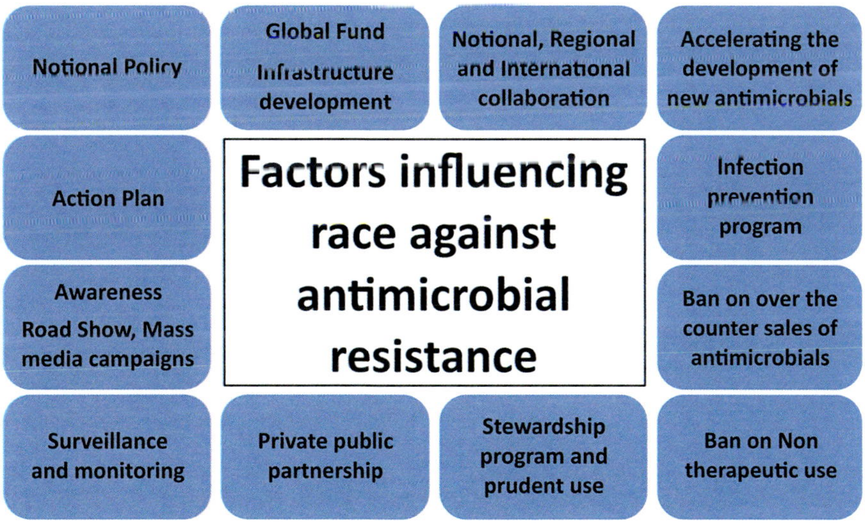

Fig. 10.5 Antimicrobial resistance—a collective responsibility (Source: Premanandh et al. 2016)

The million-dollar question then is, where does each and every one of us come into the story? In which of these niches do you fit? What role do you see yourself playing? Clearly the responsibility for eradicating antimicrobial resistance does not rest only on government officials, doctors, and healthcare workers. It is also not just limited to researchers, pharmacists, or even members of the legislative arm who create the laws, nor to educators committed to raising a growing generation of people who understand the critical role antibiotic stewardship plays but also to veterinary doctors, agricultural scientists, farmers, community health workers, and quality assurance officers who monitor effluents from various industries.

What then can we do to achieve SDG 3?

- Every one of us can walk in an increased awareness of our personal responsibility. As much as it lies on us, avoid taking unprescribed antibiotics or antibiotics for cold. And commit to completing the prescribed dosage. Don't stop the prescription simply because you feel better.
- Be an advocate, educating the public on antimicrobial resistance. Ask questions to get your neighbors and friends thinking on antimicrobial resistance and how to curb its spread. Disseminate the news for proper stewardship.
- Professionally, all hands must commit to curbing antimicrobial resistance.
- Collectively, outreaches can be organized to spread the news to both peers and upcoming generations.

10.5.1 Role of Healthcare Personnel

Historically, hospital environments have been typically noted as the melting point both for the development and spread of antimicrobial resistance due to the high consumption of antimicrobial agents in these environs. Hence it is important that key control measures are instituted in these places, and everyone has a key role to play. During the fight to curb the development and spread of the notorious methicillin-resistant *Staphylococcus aureus*, the United Kingdom instituted universal screening of incoming inpatients using nasal swabs. This was to inform patient placement and isolation to prevent the spread to other patients (Wilcox 2008). While this might not be cost-effective and feasible especially in developing countries, screening can be targeted at certain patients such as patients going in for elective surgery as part of their pre-surgery work-up and preparations.

It is also important to have a functional antimicrobial stewardship unit in hospitals that regularly audit the prescription and dispensing of antimicrobial agents, particularly controlled drugs and drugs of last resort usually reserved for critical cases. Hospitals should, as part of their clinical governance, have an antimicrobial prescribing policy in place, where guidance on empirical antimicrobials and subsequent change to narrow-spectrum antimicrobials are stated. Such a policy should also address regular susceptibility testing to monitor changing trends in susceptibility patterns and ensure that empirical treatments are still relevant. Good microbiology practices using standard tools and equipment should be promoted while improvement in the quality of antimicrobials and its supply chain should be encouraged (Rajesh et al. 2014).

There is also the highly important role of an infection prevention and control (IPC) unit in the hospital. Such a unit will oversee antibiotic surveillance in the facility and provide policy and guidelines in support of antibiotic stewardship. Also, the IPC unit is tasked with implementing hygiene and infection control practices and training of healthcare workers on their roles in preventing cross contamination and the spread of drug-resistant pathogens within hospital settings.

10.5.2 Role of Pharmacists and Pharmacy Services

Pharmacists are custodians of drugs, poisons, and medicines including antimicrobials. They are medicines experts. Pharmacists have a pivotal role in the extent and spread of antimicrobial resistance both in the hospital and community settings. The integrity of the drug supply chain is assured by pharmacists while they ensure the supply of quality medications because counterfeit drugs are not only therapeutically ineffective, they are also toxic. Pharmacists interpret prescriptions and render advice on medication needs; drug use; adverse effects; and possible drug interaction with other drugs, foods, and drinks (FIP 2015). Community pharmacists have the opportunity to engage patients in counseling and consumption of antimicrobials, dispense drugs, collaborate with healthcare professionals, discourage patients from using antibiotics without prescription, and effective drug medication and management to combat antimicrobial resistance (Howard et al. 2013; FIP 2015). Additionally, they can engage an entire community via health promotion campaigns on infection prevention, proper hygiene practices, and isolation of sick (infected) relatives. The role of the pharmacist in antimicrobial stewardship has evolved significantly since the beginning of the twenty-first century. In developed countries such as the USA and the UK, pharmacists are involved in AMS programs which has led to the reduction of antibiotic consumption, antibiotic cost, and mortality (Barlam et al. 2016).

10.5.3 Role of Drugstores

Drugstores are also known as chemist or pharmacy where over-the counter drugs are sold. Most times in some countries these drugstores do not have a resident pharmacist, those involved in selling these drugs do not have the knowledge of what AMR is and yet serve a key role in the service they provide to the general population. The drugstore owners can combat AMR by educating the public on the appropriate use of antimicrobials when they come to the drugstores for purchase (Hart and Kariuki 1998). They can only do this if they are knowledgeable enough or trained. Dispensers of antimicrobials should insist on a test result for diagnosis or doctor's prescription, because self-medication is a common practice in developing countries (Ocan et al. 2015). The quest for wealth and market competition has made many dispensers of antimicrobials prescribe what is not recommended for the patient and also purchase substandard drugs, and this should be severely discouraged. Thus, it is important to set up regulatory agencies and policies that can supervise the drugstores.

10.5.4 Role of the Public

The general public could play a key role in combating antimicrobial resistance. The practice of irrational use, misuse, and abuse of antibiotics among the public at the community level cannot be over-emphasized. This is commonly observed in developing and under-developed countries where access to healthcare services may not be affordable while antibiotics can be obtained over-the-counter without a clinician's prescription. In such a setting, there is an overall poor compliance to antibiotic regimen, self-medication practices, and other poor antibiotic-taking behavior. A recent survey carried out in Nigeria covering the six geopolitical zones reports that 31.3% of people who had used antibiotics in a six-month timeframe did not have a prescription, with 26.1% not seeing the need to complete the dose (Chukwu et al. 2020). Imagine a different scenario than this. Imagine whereby just 80% of the population is aware of the scourge of antimicrobial resistance and its effect on the human population and stands up and says, "it ends with me!" Imagine even half of this population serving as advocates for the control of antimicrobial resistance. What a victory would be had over antimicrobial resistance. The public obviously has a key role to play. This can simply begin with an individual only using antibiotics specifically prescribed for that individual. It would involve the avoidance of unprescribed antibiotics or not using antibiotics prescribed for someone else with the same "medical condition." A key step in this fight would be not taking antibiotics for viral infections and common colds. "If it wasn't prescribed for me, I wouldn't take it" should be the slogan of the day. Antibiotics are not hand-medowns; sharing is prohibited. In addition to this, antibiotics would have to be taken as prescribed and instructions judiciously followed. No matter how tempting it is to discontinue the dose with the observation of improvements in the condition, doses must be completed as this is a key step in the selection of antimicrobial-resistant strains and therefore a major "no-no." And finally, good, old, community policing: spread the news, remind your neighbors, and ask your pharmacist. Together, this scourge can be kicked out.

10.5.5 Role of the Agriculture Sector

One Health approach necessitates that the fight against antimicrobial resistance cannot take place by focusing on humans in isolation without taking into consideration the key role played by both the livestock sector and the environment. The major driving force of antimicrobial resistance in the livestock sector is the use of antibiotics for treatment, prophylaxis, and as growth factors (Fig. 10.6). Studies have reported that in the USA, the use of antimicrobial agents in food animals accounted for up to 80% of total antimicrobial use in a single year (Sharma et al. 2018). Therefore, control in this sector would be predominantly aimed at proper diagnosis, prescribing, and the judicious use of antibiotics in the livestock industry. It would also involve a clear strategy for monitoring the development and spread of resistance determinants and the avoidance of last-line resort drugs used by humans. In order to achieve this, however, a significant level of public engagement is necessary in order to sensitize key stakeholders on the implications of this seemingly "innocuous" use on human health.

10.5.6 Role of Educators

One key factor in this continuous fight against antimicrobial resistance is knowledge dissemination. For everyone to be on the same page, actively working together to control and reduce antimicrobial resistance not just locally, but globally, education is imperative. Educators therefore

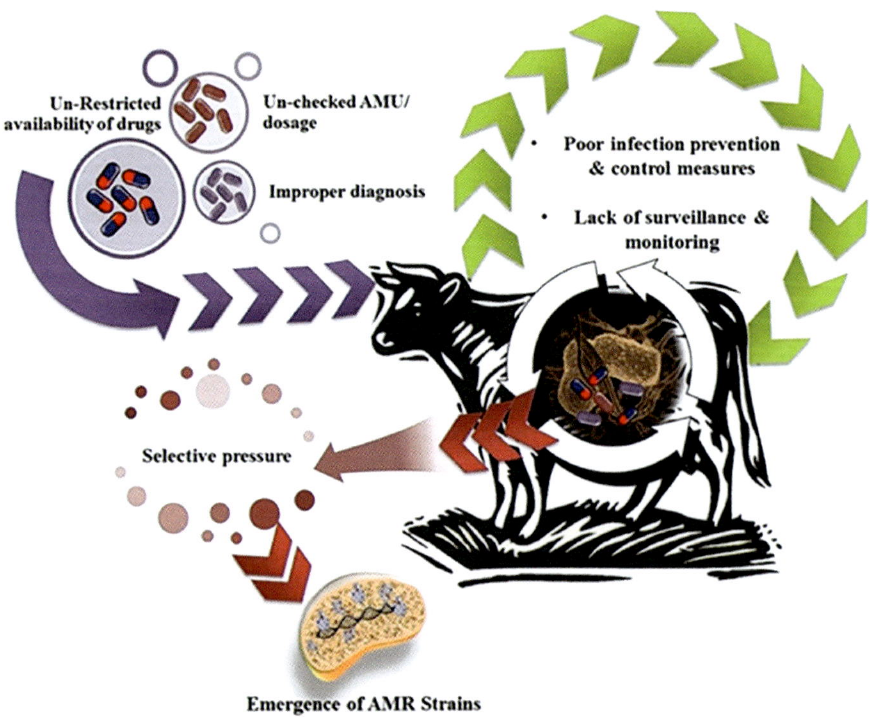

Fig. 10.6 Potential sources of antimicrobial resistance in animals (Sharma et al. 2018)

have a key role to play especially as this cuts across the different stakeholders. It is important to properly package the information in order to directly appeal to the specific target audience. One way to do this might be to organize train-the-trainer programs designed to inform key people in the target groups with the ability to cascade this information down to all members of the group. This could involve training by the antimicrobial stewardship group or the public health physicians to the various cadre of hospital personnel with a specific focus on what each group can do to join the war. Additionally, in-patients can be educated via flyers or posters strategically situated around healthcare settings to sensitize the population. Key target groups like women enrolled for antenatal care can then be directly given a talk on their role in combating antimicrobial resistance knowing that these would further disseminate the information to their families and communities. The effects of a number of non-governmental action groups with AMR as their focus have been observed with one of these (Community Health Awareness Network based in Enugu, Nigeria) noted to be taking the AMR awareness campaign to rural areas in their locale. A higher level of proliferation of these should be encouraged with interworking networks to enable easy sharing of knowledge. This education can take the form of radio jingles and sensitization of school children in a bid to create early awareness. And taking this a step further, key industries can be sensitized in order to incorporate AMR awareness as part of their Social Responsibility. The 18th to 24th November of each year has been set apart as the World Antimicrobial Awareness Week by the WHO and would provide a good opportunity for massive awareness campaigns not just in urban areas but also in rural areas.

So, the question today is: "Will you commit to playing your own role?"

10.6 Conclusion

Antimicrobial resistance is an increasingly present threat with dire consequences to the human race. The fight to stop both the development and spread of these resistant strains is on but needs to be escalated and must be won on all fronts, as a loss at one front will translate to a loss of the entire war. It is our health, our responsibility and each one of us owes it to generations to come to leave the world better than we met it. The scourge of antimicrobial resistance can and will be eradicated. The onus is ours to play our role. Each of us fighting in our little way, with the tools we have, to end this growing trend of drug resistance and change the current narrative.

References

Abubakar U, Sulaiman SAS. Prevalence, trend and antimicrobial susceptibility of methicillin resistant *Staphylococcus aureus* in Nigeria: a systematic review. J Infect Public Health. 2018;11(6):763–70.

Barlam TF, Cosgrove SE, Abbo LM, MacDougall C, Schuetz AN, Septimus EJ, et al. Implementing an antibiotic stewardship program: guidelines by the Infectious Diseases Society of America and the Society for Healthcare Epidemiology of America. Clin Infect Dis. 2016;62(10):e51–77.

Bernabe KJ, Langendorf C, Ford N, Ronat JB, Murphy RA. Antimicrobial resistance in West Africa: a systematic review and meta-analysis. Int J Antimicrob Agents. 2017;50(5):629–39.

Booton RD, Meeyai A, Alhusein N, Buller H, Feil E, Lambert H, Mongkolsuk S, Pitchforth E, Reyher KK, Sakcamduang W, Satayavivad J. One health drivers of antibacterial resistance: quantifying the relative impacts of human, animal and environmental use and transmission. One Health. 2021;12:100220.

Cassini A, Högberg LD, Plachouras D, Quattrocchi A, Hoxha A, Simonsen GS, Colomb-Cotinat M, Kretzschmar ME, Devleesschauwer B, Cecchini M, Ouakrim DA. Attributable deaths and disability-adjusted life-years caused by infections with antibiotic-resistant bacteria in the EU and the European economic area in 2015: a population-level modelling analysis. Lancet Infect Dis. 2019;19(1):56–66.

Chukwu EE, Oladele DA, Awoderu OB, Afocha EE, Lawal RG, Abdus-Salam I, Ogunsola FT, Audu RA. A national survey of public awareness of antimicrobial resistance in Nigeria. Antimicrob Resist Infect Control. 2020;9(72):72.

Dadgostar P. Antimicrobial resistance: implications and costs. Infect Drug Resist. 2019;12:3903.

Davey P, Brown E, Charani E, Fenelon L, Gould IM, Ramsay CR, et al. Interventions to improve antibiotic prescribing practices for hospital inpatients. Cochrane Database Syst. 2013;9(2):CD003543.

Davies J, Davies D. Origin and evolution of antibiotic resistance. Microbiol Mol Biol Rev. 2010;74(3):417–33.

Dewachter L, Fauvart M, Michiels J. Bacterial heterogeneity and antibiotic survival: understanding and combatting persistence and heteroresistance. Mol Cell. 2019;76(2):255–67.

Dhingra S, Rahman NAA, Peile E, Rahman M, Sartelli M, Hassali MA, et al. Microbial resistance movements: an overview of global public health threats posed by antimicrobial resistance, and how best to counter. Front Public Health. 2020;8:531.

Filice GA, Nyman JA, Lexau C, Lees CH, Bockstedt LA, Como-Sabetti K, Lesher LJ, Lynfield R. Excess costs and utilization associated with methicillin resistance for patients with *Staphylococcus aureus* infection. Infect Control Hospital Epidemiol. 2010;31(4):365–73.

Founou RC, Founou LL, Essack SY. Clinical and economic impact of antibiotic resistance in developing countries: a systematic review and meta-analysis. PLoS One. 2017;12(12):e0189621. https://doi.org/10.1371/journal.pone.0189621.

Hart CA, Kariuki S. Antimicrobial resistance in developing countries. BMJ (Clinical research ed). 1998;317(7159):647–50. https://doi.org/10.1136/bmj.317.7159.647.

Hernando-Amado S, Coque TM, Baquero F, Martínez JL. Antibiotic resistance: moving from individual health norms to social norms in one health and global health. Front Microbiol. 2020;11:1914.

Howard P, Ashiru-Oredope D, Gilchrist M. Time for pharmacy to unite in the fight against antimicrobial resistance. Pharm J. 2013;291(537–8):53.

International Pharmaceutical Federation (FIP). Fighting antimicrobial resistance: the contribution of pharmacists. The Hague: International Pharmaceutical Federation; 2015.

Lee AS, De Lencastre H, Garau J, Kluytmans J, Malhotra-Kumar S, Peschel A, Harbarth S. Methicillin-resistant *Staphylococcus aureus*. Nat Rev Dis Primers. 2018;4(1):1–23.

Mulani MS, Kamble EE, Kumkar SN, Tawre MS, Pardesi KR. Emerging strategies to combat ESKAPE pathogens in the era of antimicrobial resistance: a review. Front Microbiol. 2019;10:539.

Mutuku C, Gazdag Z, Melegh S. Occurrence of antibiotics and bacterial resistance genes in wastewater: resistance mechanisms and antimicrobial resistance control approaches. World J Microbiol Biotechnol. 2022;38(9):152.

Ocan M, Obuku EA, Bwanga F, Akena D, Richard S, Ogwal-Okeng J, Obua C. Household antimicrobial self-medication: a systematic review and meta-analysis of the burden, risk factors and outcomes in developing countries. BMC Public Health. 2015;15(1):1–11.

O'Neill J. Tackling drug-resistant infections globally: final report and recommendations; 2016. Available online at: https://amr-review.org/sites/default/files/160518_Final%20paper_with%20cover.pdf . Last Accessed 27 Sept 2021.

Peterson E, Kaur P. Antibiotic resistance mechanisms in bacteria: relationships between resistance determinants of antibiotic producers, environmental bacteria, and clinical pathogens. Front Microbiol. 2018;9:2928.

Premanandh J, Samara BS, Mazen AN. Race against antimicrobial resistance requires coordinated action–an overview. Front Microbiol. 2016;6:1536.

Qu J, Huang Y, Lv X. Crisis of antimicrobial resistance in China: now and the future. Frontiers in Microbiology 2019;10:2240.

Rajesh RU, Gurdeep SK, Vijay MK, Onkar CS. Strategies to combat antimicrobial resistance. J Clin Diagn Res. 2014;8(7):ME01-ME04.

Schuts EC, Hulscher ME, Mouton JW, Verduin CM, Stuart JW, Overdiek HW, van der Linden PD, Natsch S, Hertogh CM, Wolfs TF, Schouten JA. Current evidence on hospital antimicrobial stewardship objectives: a systematic review and meta-analysis. Lancet Infect Dis. 2016;16(7):847–56.

Sharma C, Rokana N, Chandra M, Singh BP, Gulhane RD, Gill JPS, Ray P, Puniya AK, Panwar H. Antimicrobial resistance: its surveillance, impact, and alternative management strategies in dairy animals. Front Veterinary Sci. 2018;4:237.

Wada Y, Harun AB, Yean CY, Zaidah AR. Vancomycin-resistant enterococci (VRE) in Nigeria: the first systematic review and meta-analysis. Antibiotics (Basel, Switzerland). 2020;9(9):565.

Wall S. Prevention of antibiotic resistance–an epidemiological scoping review to identify research categories and knowledge gaps. Glob Health Action. 2019;12(sup1):1756191.

Wilcox MH. Screening for MRSA. Br Med J. 2008;336(7650):899–900.

World Health Organization. *Antimicrobial resistance*. Geneva: Key Facts; 2018. Available online at: https://www.who.int/news-room/fact-sheets/detail/antimicrobial-resistance . Last Accessed 27th September, 2021

11 Accessing Library Information Services in Public Health Emergencies

Helen Uzoezi Emasealu

11.1 Introduction

The focus of this chapter is to state the relevance of information in a dynamic modern world and suggest how members of the public can access information in attending to public health emergencies. Information activates and stimulates the human thought process. It is the key to effective decision-making for a meaningful and sustainable world order, be it in the present or the future. For a society to function well, information is key. Information is so important that it can be rightly described as the breath and blood of every useful human engagement. Information has become critical and central for better communication and productivity for parents, company executives, the legal profession, the entertainment industry, the economic and political policy sphere, the world of commerce, and the health sector, among others. The place of information in the new dynamic world cannot be overemphasized. Indeed, information rules the world. Information, therefore, is not only limited to the academics but also for daily functioning, and it is equally very important for healthy living, especially during public health emergencies. Kaye (1995) averred that 'Good information, improves decision-making, enhances efficiency (Nwaichi and Abbey 2015) and provides a competitive edge to the organization which knows more than the opposition'. In the same manner, the informed would always maintain good hygiene, get vaccinated, identify with precautionary measures, and stay healthy. Whereas the reverse would be the case for the uninformed. Most often, vices and undesirable behaviour exhibited by an individual are resultant of a lack of information or misinformation. Hence, information also acts positively on the behaviour of individuals and the society at large.

Information adds value such that the bearer is always in an advantageous position in society. However, not all information is correct or beneficial to the bearer, and therefore, individuals should endeavour to access credible information from verifiable sources. Consequently, this chapter on eulogizing information, its importance and assessing library information in public health emergencies, highlighted the concepts of the information cycle and misinformation.

11.1.1 Information Cycle

Information is commonly defined as processed data. This simply means that information is an interpretation or answer to a perceived problem, event, or idea. Hence, information is the product of the brainchild of the originator. Information can be new or built upon previously existing ones; however, new information is subjected to

H. U. Emasealu (✉)
Department of Library and Information Science, University of Harcourt, Choba, Rivers State, Nigeria
e-mail: helen.emasealu@uniport.edu.ng

E. O. Nwaichi (ed.), *Modernity in Health and Disease Diagnosis: The Account from STEM Women*, Sustainable Development Goals Series, https://doi.org/10.1007/978-3-031-34963-8_11

credibility assessment through peer-review processes or against established criterion before they are accepted or generalized.

Information is being produced daily. Consciously and unconsciously, people provide data for information and feedback. There is exponential growth in the production and use of information. This led to the concept of prosumer in information processes where information is being produced and used simultaneously. Information is produced and communicated to the public or targeted audience. The audience receives and synthesizes the information, translating it into practical use as well as providing judgemental data or giving value based on the credibility of the information received. These are transmitted back to the originator in the form of passive feedback.

The library is the social institution for credible information established to professionally provide information services to the public. The librarians are professionally trained to assess the credibility of information before acquisition for public consumption. The primary role of the library is to acquire and catalogue information (grouping information in an organized manner), preserve human knowledge, disseminate information, and provide information services. Therefore, the library remains a veritable information source for members of society. Haider (2020) remarked,

> Who needs a librarian and cataloguer when you have google and internet? Well, who needs a teacher when you have Wikipedia? And who needs a doctor when you have WebMed?... Just as the Wikipedia doesn't replace the teacher, and WebMed doesn't replace the doctor, in the same way, google search and internet doesn't replace the librarian and cataloguer.

This supports the argument that information outside the confines of the library maybe misinforming. This is because, unlike other information centres, the library is the only social institution with the sole responsibility to seek and acquire credible information in all formats for dissemination.

11.1.2 Misinformation

Emasealu and Ezeonye (2022) have observed that there are information-vulnerable members of society. This group of people are exposed to the dangers of inaccurate information, which are in frequent circulation. This can come at a terrible cost for the affected individual in particular and the society at large. Public health would suffer adversely when exposed to inaccurate information or misinformation. Misinformation is that information which is baseless and carries falsehoods, rumours, hearsays, and unsubstantiated accounts emanating from the sheer imagination of the information provider. People who are oblivious of their information needs or the usefulness of information are often at the peril of wrong information. Accordingly, Emasealu and Ezeonye (2022) categorized information users into:

1. The star information users—are those who appreciate the importance of information and are in constant search to improve upon their knowledge base through the acquisition of credible information. This group according to the authors are intrinsically motivated to create, use, and seek information.
2. The clique (information segregationists)—is made of individuals who placed self-imposed restrictions on the type of information they seek. This is usually restricted by profession, career, business, or academic focus. The authors explained that those in this category are limited 'in the upwardly mobile contemporary world'.
3. The rejectee (information-static)—are the group of persons that are not progressive with information acquisition as they are only open to information received during formal training, with total disregard to information as an ingredient for daily functioning.
4. The isolate (information-vulnerable)—are classified as those who are oblivious of their information need and do not benefit from the occasion of information explosion but rather depend on hearsay as sole and credible sources of information. The authors explained that 'it is in this group that the effects of fake news, rumours, conspiracy theories, and the like are strongly felt'.

People Perish for Lack of Knowledge. The biblical salvation call 'My people are destroyed for lack of knowledge. Because you have rejected knowledge, I also will reject you…' (Hosea 4:6, n.d) amplifies the importance of knowledge, bearing in mind that knowledge thrives, principally, on received useful information. An example that readily presents itself on how people can be destroyed for lack of knowledge is that of those who do not believe that COVID-19 exists simply because they are not locked in for emergency purposes. Such persons do not realize the severity of COVID-19 as it is being portrayed. This attitude is attributed to a lack of information or even worse, misinformation. The implication of this is that many would perish for lack of knowledge, knowledge being a derivative of accurate information. Misinformation often poses a greater risk in most cases. Accordingly, the UN strategic development goal 3 of good health and well-being objectifies 'strengthening the capacity of all countries, in particular developing countries, for early warning, risk reduction and management of national and global risks' (World Health Organisation 2016). In the same vein, the United Nations' definition of health and well-being is not limited to the absence of ailment but also freeness from emotional disturbances. Accordingly, information has the capacity to alter emotions and decide the course of action for individuals; however, the consequence of such action is the determinant on the correctness or wrongness of such information. Hence, the ABC model of action by Ellis, a prominent psychologist, means that the information available to a person plays the sole role of the definition such person ascribes to situations around him/her which further decides the action to follow. Arguably, such a situation refers to health emergencies and misinformation, as individuals may be wrongly informed during health emergencies, leading to a disastrous situation than before.

Sell et al. (2020) have explained that most information circulated on social media during the Ebola pandemic in the disguise of health information was intended for politics, jokes, and fear-induction among others which contained, partially, health-related information. Such information is weaponized and could lead to unwanted outcomes as they resonate differently with individual beliefs, as it is not health intended. Also, such misinformation has been cited to fuel vaccine hesitancy among the public (i.e. as witnessed with polio vaccination) and impede recovery efforts. The viral disease impacts human social life with certain changes, and adjusting well to the new realities requires accurate and timely information. For efficient and effective public response and recovery, therefore, effective communication is a deciding key factor. It is observed that fear, uncertainty, illiteracy, and information needs may increase the chances for the propagation of health misinformation. Information provision services in the library do not stop even in public health emergencies as users can utilize communication technologies to access the services. Therefore, the belief that librarians become inaccessible during health emergencies is fallacious to achieve the expectations of SDG 3.

Misinformation is detrimental to human life and societal progression, more so, misinformation during public health crises, for instance, can have adverse effects because it can lead people astray by making wrong decisions that would aggravate an already existing challenge. Health Misinformation (HM) has been identified to pose disastrous effects that could be irreversible, such as death. Kolluri and Murthy (2021) see health misinformation as a serious threat to public health, that is capable of causing confusion, sowing mistrust, causing harm to people's health, and undermining public health efforts. The 2019 outbreak of the COVID-19 pandemic generated widespread fear, changed the lifestyle of many, and posed health emergencies. Libraries, among other social institutions, were closed down and movement was restricted to curtail the spread of the pandemic. The need for public health information became increasingly important to be updated on the progress made in mitigating the effects of the coronavirus.

The history of mankind has constantly been challenged by live threatening occurrences. Cases of pandemics have been traced from 1346 with the case of the black death, the Spanish flu, the plague of Justinian, HIV/AIDS, to the most recent being COVID-19. Hence, life is constantly faced with strange variants of diseases that require careful examination for a vaccine which requires some amount of time. While the experiment lasts, the society needs every available credible information about how to stay safe and alive. This need for information could be overwhelming, leading to individuals engaging in information dissemination whereas they may not be professionally equipped with information assessment skill sets. This and the emergence of communication technologies have constituted challenges in information management and the dissemination role of librarians. Umeozor (2020) opined that the ubiquity of the internet and web connectivity has created the illusion of self-sufficiency in individuals with access to the internet in information retrieval and use. However, libraries and librarians are transforming the philosophy of their services from material-based to client-based. This is to provide need-based library and information services to ensure patron satisfaction. Hence, librarians are serving as mediators (brokers) between users and the myriad of information sources.

11.2 The Challenge of Information Management in the Modern World

In the twenty-first century, there has been an exponential rate of information creation, which is generated daily and at a very fast rate. In the same century, librarians are not the only agents of information dissemination. Others are relatives, friends, teachers, journalists, organizations such as trade associations, business firms, labour unions, professional societies, educational institutions, publications, career centres, internet service providers such as website application providers, blogs, articles, art exhibitions, and displays in the form of films and videos among others.

Furthermore, there is a mutated need for information. The search for information is becoming increasingly difficult to keep up with, as there are many sources of information, yet the needed information is scarce or few. In most developing countries, the library profession is not considered prestigious in the hierarchy of professional courses, yet the society demands the few library professionals to effectively satisfy their various information needs.

Also, the increasing transformation and development of information communication technology is making it difficult for librarians to keep up, thereby, causing unplanned obsolesce of most library practices. Most librarians were trained with a curriculum older than the current situation of events. They are therefore tasked with acquiring technical skills in using the technologies of communication in rendering information services that are not bound by the four walls of their libraries.

There is also the issue of funding. Libraries are advocating the adoption of communication technologies. However, they are facing funding problems in acquiring the equipment. In most cases, where the equipment is procured, the use and maintenance culture become a problem as librarians need training and retraining in keeping up with these tools of communication. All these are achieved with adequate funding. Hence, the reoccurring issue of inadequate funding in libraries continues to clog the wheels of library and information services and management in the twenty-first century.

11.3 Importance of Librarianship

With the myriad sources of information provision, there is a challenge in information handling in the modern world. However, whereas technology capacitates just anybody with access and the ability to disseminate such information, it may not equip everyone with the professional skill set of information sourcing, verification for authenticity, identification of information needs, and effective dissemination. This is the job of the professional librarian and herein lies the importance of the librarian.

In the traditional sense, the librarian is the information specialist that acquires, manages, preserves, organizes, and disseminates information resources for the benefit of the information end user. However, with the rapid rate of development in the information society in the recent past, the role of the librarian has become more challenging as it now includes mediator, change agent, propagator of knowledge, and perhaps, most importantly, the first responder that acts as the catalyst between the modern information superhighway and the information user. In a modern world characterized by a superhighway of information flow, the librarian is no longer a mere custodian of information. To underscore this point, Kwanya et al. (2015) describe librarians as apo-mediators. This means that the librarian is no longer the professional who selects and acquires information resources in all formats, sits behind the desk checking identity cards, and admits users of information into the library building. Librarianship has gone beyond this.

The modern librarian has evolved to become a professional who is vastly knowledgeable about his environment and is available to offer direct professional assistance in general counselling on information handling which includes understanding information needs, identifying verified sources of information, retrieving the needed information, and ensuring adequate use of information. The library is an information-neutral institution. Therefore, the library must remain a professional place for factual evidence of events in human society. In the same vein, the mantra of the library at all times must be getting the right information, at the right time, in sufficient quantity to the right persons, and help distinguish between facts and myths.

It is noteworthy that in acknowledgement of the power of information, the government and the affluent owned libraries in ancient Greek and Roman societies. In present times, information is provided free of charge by librarians for information users whether in a library building or within a community. The library must be seen as being more than a physical building with people, books, and computers. The library encompasses the creation of knowledge and the dissemination of information for targeted objectives. Beyond the physical structure, the library boasts the resourcefulness to operate remotely in promoting information access and use. The library provides general information and specific information needs.

(a) General information need—Information behaviour subsumes expressed and unexpressed information need. Most persons are not aware of their current need of information. The general information need includes information classified by discipline or intended to inform the general public without particular reference to any individual. For example, information as broadcasted news, posters, announcements, and fiction among others. These classes of information are informing on general issues or interests. This type of information usually addresses unexpressed information needs.
(b) Specific information need—Specific information need is seen as marked with intent. It is characterized by the desire to seek and use information on an identified problem situation or need and must be deliberate and strategic (Nwaichi and Abbey 2015). For example, searching for information on how to use a particular phone function, cook a meal, or make a particular design of an object. It is specific because they could be embedded in a body of general information, but the importance is centred on a particular part of the whole information. Another example is seeking for information about a particular function of a computer from a complete computer manual. In libraries, reference librarians provide professional assistance in the search, retrieval, use, and dissemination of specific information services.

11.4 Accessing Library Information in Public Health Emergencies

The factors that militate against the attainment of effective librarianship in public health emergencies include rank illiteracy, abysmal ignorance, communication gaps, poor handling of information resources, and lack of synergy between public health officers and librarians. It is a sad fact that human existence has always been and will continue to be challenged by various viral outbreaks. The world must get ready for such future occurrences by developing strategies to effectively combat them. Libraries, as social institutions of information also, should strategize and adopt meaningful means of disseminating useful information during such outbreaks.

In emergencies occasioned by disease outbreaks, libraries should be engaged for information needs, retrieval, and use. The three approaches canvassed to bridge an existing gap in accessing library services in public health emergencies are as follows:

1. Know your library and librarian (reference librarian).
2. Community collaborations (town and gown).
3. Library subscriptions (websites and mobile applications).

11.4.1 Know Your Library and Librarian (Reference Librarian)

Libraries are categorized by their types, which is a determinant of the services they provide. It is important to note that the library does not segregate in the provision of services to individuals; however, the knowledge of the types is crucial in determining the appropriate library (Table 11.1) to seek particular information. The library user must know their environment sufficiently so that in cases of public health emergencies, such as Ebola, Covid-19, monkeypox, cholera, et cetera, they can know the nature of the emergency so that useful information can be given to the librarian to source for the most relevant information.

Therefore, the user should be acquainted with the type of libraries and area of information service coverage, register with the library, identify with the reference librarian and refer-

Table 11.1 Library types

S/No	Type	Information service	Clientele
1	Academic library	Research and academic programmes supporting information services to universities, polytechnics, monotechnic, and every other type of tertiary institution.	Academicians and researchers
2	School library	Provides information services that foster learning and promotes literacy in school children.	Secondary and primary school children
3	Public library	Provides information resources to promote literacy and provide access to government publications and documents. This is subsumed under the National Library which serves as a country's repository.	
4	Special library	Focuses on a specialized type of information tailored towards a particular person, object, situation, phenomenon, or event, for example, petroleum.	Organizational bodies
5	National library	The library of other libraries. The National Library keeps records of all publications in a country and helps gives national identity to every book published by authors of a country by assigning international standard book number (ISBN) and international standard serial number (ISSN). It also keeps a copy of every publication within a country over the years.	National publishers and other libraries
Further grouping			
1	Community/mobile library/community information centres	This is a need-based library. It is mostly a makeshift public library usually set up to serve immediate information needs as identified in a particular area. It does not dwell on a particular information service but on situational information needs. An example is a mobile library in communities on prevailing health conditions in the location.	The population of where it is situated

ence resources available, and indicate their information interest with the strategic selective dissemination of information (SSDI) service of the library. Following these steps helps the reference librarian to provide specific information of interest to individual patrons even in remote locations. The SSDI service helps the users to indicate their information preferences with which the reference librarians notify them of the existing and new arrival of information based on their personal preferences. Thus, reference librarians of every library can effectively operate on a round-the-clock basis to provide crucial information, especially during public health emergencies.

11.4.2 Community Collaborations (Town and Gown)

The library does not function in isolation. The users of information are the major stakeholders. The 'town and gown' initiative is the brainchild of intentional community development organized in universities for host communities. While libraries are limited in their scope, they provide information services aimed at societal development. This involves active participation in community development activities which over the years have included such activities as creating pathways, sanitation, health outreach, sensitization, and orientation among others. Additionally, libraries have keyed in to provide information resource centres. These centres bridge the information gaps between the universities and their host communities. Community members should leverage on this to access health information that is mostly disseminated in local languages than depend on hearsay.

Importantly, community libraries are also skill acquisition centres. These libraries create makers space where users are taught varied types of skills. Community libraries are established with a focus on the identified information and skill needs of a community as a means of instilling consciousness and functional citizenry among community members.

11.4.3 Library Subscriptions (Websites and Mobile Applications)

The exponential growth in technologies of communication offers people more ways to connect, communicate, and share information and has, thus, become the primary mode of communication because of its speed and versatility. However, the relevant questions are:

- Is the information adequate (scope)?
- Where is the information from (source)?
- Who is the authority of the information (author/publisher)?
- Can the information solve the specific need (usefulness)?

Libraries have consciously adopted and explored communication technologies to promote information services and offer virtual services such as interlibrary loans, reference services, outreach services, prompt notification of library development, new arrivals/acquisitions, institutional repository updates, library orientation, user education, and language translation (Emasealu and Umeozor 2018). Library subscription will enable the library to keep a record of its active users and subsequently, dispatch information resources/documents to their respective addresses. Consequently, users should familiarize themselves with the available web-based automated library information services through the library's social media accounts and handles, websites, homepages, and mobile applications to access unabridged information services without the intervening presence of misinformation.

11.5 Conclusion

This work has examined the role and importance of communication and established that information is key to effective decision-making for meaningful and sustainable world order. The work also established that health misinformation can lead to fear, vaccine hesitancy, death, and ineffective public recovery that comes at a terrible cost for society. The challenges of information handling in the modern world such as the exponential growth of media of communication and associated technologies have been cited as fuelling the wheels of misinformation. The work espoused the importance of the librarian as the professional information manager, citing the library as an information-neutral entity. Consequently, the chapter enumerated ways library information services can be accessed during public health emergencies to ensure improved access and dissemination of verifiable information aimed at the promotion of human society.

11.6 Recommendations

- The library patron should be able to seek and share only credible information.
- The end user must be ready to circulate and pass on this useful information to other members of the community rather than hoarding it to avoid misinformation circulation.
- The librarian must know the structure of the community and how to reach out to leaders of the community (market women leaders, union leaders et cetera) so as to permeate the community with information.
- Regular education in collaboration with authorities (schools, professional bodies, religious leaders) to sensitize the people and create awareness on public health issues.
- The patron should also seek relevant information on climatic factors and how this information can impact public health.
- Adequate funding of libraries.
- Training and retraining of librarians.

References

Emasealu HU, Ezeonye SN. The first responder librarian as a guide to the information-vulnerable members of society. J Library Serv Technol. 2022;3(2):112–8. https://credencepressltd.com/journal/uploads/archive/202216522623173613722347.pdf

Emasealu HU, Umeozor SU. Application of social media in the promotion of research activities: librarians as catalysts. Library Inform Sci Digest. 2018;11(2):22–35.

Haider S. Who needs a librarian and cataloguer when you have google and internet. (2020). https://www.librarianshipstudies.com/2018/07/who-needs-a-librarian-and-cataloger-when-you-have-google-and-internet.html

Hosea (Chap. 4.6), (n.d) KJV. https://www.thebible.com. Accessed on 12/9/2022.

Kaye D. The importance of information. Manag Decis. 1995;33(5):5–12. https://doi.org/10.1108/EUM0000000003897.

Kolluri NL, Murthy D. CoVerifi: a COVID-19 news verification system. Online Soc Netw Media. 2021;22:100123., ISSN 2468-6964. https://doi.org/10.1016/j.osnem.2021.100123.

Kwanya T, Stiwell C, Underwood P. Library 3.0: intelligence libraries and apomediation. Kidlington: Chandos Publishing; 2015.

Nwaichi EO, Abbey BW. Essentials of scientific communication in a productive research. J Sci Res Rep. 2015;7(7):525–31.

Sell TK, Hosangadi D, Trotochaud M. Misinformation and the US Ebola communication crisis: Analysing the veracity and content of social media messages related to a fear-inducing infectious disease outbreak. BMC Public Health. 2020;20:550. https://doi.org/10.1186/s12889-020-08697-3.

Umeozor SN. Information retrieval: a communication process in the 21st century library. Int J Knowl Content Develop Technol. 2020;10(2):7–18.

World Health Organization. Regional Office for the Western Pacific. (2016). Sustainable development goals (SDGs): Goal 3. Target 3.d: Strengthen the capacity of all countries, in particular developing countries, for early warning, risk reduction and management of national and global health risks [poster]. WHO Regional Office for the Western Pacific. https://apps.who.int/iris/handle/10665/208291

12 Human Gut Microbiome: Its Role in Health and Development

Obakpororo Ejiro Agbagwa

12.1 Introduction

Scientists estimate that about 100 trillion microbes are found in the human body and that most of these organisms are within the gastrointestinal tract (GIT). The microbes living within the human intestine are referred to as the gut microbiota. The human GIT is one of the largest interfaces between the host, environmental factors, and antigens in the human body. Most diets with numerous microbes go through the GIT. The species, their diversity, and total genetic makeup of these organisms found within the GIT (intestines) are referred to as the gut microbiome (GM). Most of these organisms are beneficial to human health while others can cause disease and malfunctioning of the body [Integrative HMP (iHMP) Research Network Consortium]. They have the ability to affect the physiologic functions of the body (Bengmark 1998). A healthy gut consists of the phyla *Bacteroidetes*, *Firmicutes*, *Actinobacteria*, and *Proteobacteria* in abundance, while *Akkermansia*, *Bifidobacterium*, and *Escherichia* are less abundant. A healthy gut has the ability to keep and maintain the contents of the gut by preventing the escape of undigested food, toxins, and harmful metabolites from escaping via the bloodstream. It is important that there is a balance between the commensal and pathogenic microbes within the gut so as to maintain good health, development of the human body, and disease prevention. The imbalance of the human GM is called dysbiosis. When the gut barrier is weakened it allows the passage of small particles comprising of bacteria and food particles into the bloodstream resulting in a leaky gut. Dysbiosis is linked to inflammatory bowel disease, irritable bowel syndrome, allergy, obesity, colon cancer, mental illness, and other diseases (Qin et al. 2010; Thursby and Juge 2017). Maintaining a healthy GM will aid in preventing microbiome-related diseases, thus promoting a healthy life and well-being in line with SDG 3.

12.2 Development of the Gut Microbiome

The human GM begins its development from the utero when the GIT is exposed to microbes. Before birth, the gastrointestinal tract of a normal fetus is sterile. Microbes colonize the gut during or immediately after birth. During the first year of life, the composition of the gut microbiota is simple and is instrumental in shaping the composition of the adult gut microbiota. A study showed that the bacterial counts in an infant of one month delivered via vagina were higher than those delivered by cesarean section (Huurre et al. 2008).

O. E. Agbagwa (✉)
Department of Microbiology, University of Port Harcourt, Port Harcourt, Rivers State, Nigeria
e-mail: obakpororo.agbagwa@uniport.edu.ng

E. O. Nwaichi (ed.), *Modernity in Health and Disease Diagnosis: The Account from STEM Women*, Sustainable Development Goals Series, https://doi.org/10.1007/978-3-031-34963-8_12

The diversity of species in the fetus in the womb is low due to limited colonization with microbes (Kovatcheva-Datchary et al. 2015). After birth, the GIT diversity of species begins to increase as the infant gets in contact with microbes via the channel of birth, illness, antibiotic treatment, environmental factors, and feeding. At this stage, the GIT is characterized by lactobacilli (vaginal delivery) and *Bacteroides* genus (C-Section) (Rodriguez et al. 2015; Avershina et al. 2015). Infants born by vaginal delivery are exposed to the mother's bacteria at birth, and this influences the infant's gut bacteria and stimulates white blood cells and other components associated with the immune system (Kulas et al. 2013). The GIT of a fetus can also get colonized by bacteria through amniotic fluid that was swallowed. Previous studies have shown that bacteia can efflux releasing chemicals from a mother's gut to that of the fetus (Jiménez et al. 2008). As adulthood sets in, the diversity of species present in the GIT becomes stable and distinct with lower levels of *Bacteroides* and higher levels of *Bifidobacterium* (Hollister et al. 2015). An infant's gut microbiota undergoes stages of changes attributed to the mode of feeding and the microbiota of the mother. The mother's microbiota is influenced by environmental factors, daily diet, physiological factors, etc. The adult human GM comprises six to seven different bacteria species. The findings of a study conducted showed that 60–70% of bacterial strains present remained stable and no change was observed among the strains in the microbiome (Faith et al. 2013). As an individual ages, the diversity of microbes within the human GM declines with the state of health of the individual. The aged human GM can be influenced by dietary alterations, antibiotic usage, and physiologic changes which can influence the diversity of strains and the health of the human gut microbiome.

12.3 Roles of the Human Gut Microbiome

A healthy human GM assists in determining the health status of an individual, through interaction with the brain. It plays a major role in health promotion, development, and disease prevention in the human body. One of the several roles of the GM is metabolism by gut bacteria. The GM produces vitamins that are necessary for biotransformation. It has the ability to ferment carbohydrates especially the non-digestible fibers which produce short-chain fatty acids (SCFAs) useful to maintain a healthy gut. Examples of SCFAs are acetate, propionate, and butyrate. Acetate is the most abundant of the three and it's produced by most gut anaerobes, whereas propionate and butyrate are produced by different subsets of gut bacteria following distinct molecular pathways (Louis and Flint 2017). Acetate is also involved in the regulation of metabolism in extra-intestinal tissues including cholesterol metabolism and lipogenesis. Butyrate is an important energy source for the cells that make up the inner lining of the human colon (colonocytes). When SCFA is lacking, it can lead to microbiome-related problems such as leaky gut and local inflammation. Butyrate has been linked to several important functions such as the ability to induce apoptosis of colon cancer cells and activation of intestinal gluconeogenesis which is relevant for energy balance and diabetes. Propionate is important for glucose homeostasis as it regulates gluconeogenesis in the liver and is involved in satiety signaling (Morrison and Preston 2016; Lin and Zhang 2017).

Human GM takes part in enzyme production and digestion of food, defends against opportunistic pathogens, and supports intestinal immune function and modulation. The GM regulates the brain function through tryptophan metabolism which in turn affects human behavior (Cheolmin and Yong-Ku 2021). The gut-brain axis functions as a bidirectional communication network via the immune and endocrine system that links emotional and cognitive centers of the brain. This is regulated through the metabolism of microbiome-derived neurocompounds and tryptophan which is beneficial in nerve development. But a microbial imbalance in the human GM can lead to neuropsychiatric illnesses such as depression and schizophrenia. Stress has also been shown to influence the integrity of the gut epithelium and to alter peristalsis, secretions, and mucin production. When this occurs, the habitat of the intestinal microbiota is altered and thus promotes changes in microbial composition and metabolism of GM (Foster and McVey Neufeld 2013).

There is a link between gut microbiome and mental health. Recent studies have shown that intestinal microbiota can affect the brain's normal function and development. This was studied by comparing bacterial composition in children with ADHD and those without ADHD or autism. The researchers confirmed that the bacterial composition in children with ADHD was different from those without. Some bacteria present in the GM produce antimicrobial metabolites which enable them to inhibit harmful invaders.

12.4 Conditions Linked to Dysbiosis in the Human Gut Microbiome

Dysbiosis of the human GM can lead to irritable bowel syndrome (IBS), inflammatory bowel disease (IBD), chronic liver disease, peptic ulcer disease, *Clostridium difficile* infection, celiac disease, diverticular disease, persistent antibiotic-induced colitis, cardiovascular disease (CVD), and GI cancers. GM microbial imbalance is also linked to metabolic disorders (type 2 diabetes) and obesity and those associated with the brain such as autism, anxiety/depression, narcolepsy, and dementia. IBD are associated with alterations in the gut microbiota, though it is not clear if microbial changes contribute to disease pathogenesis or develop as a result of local inflammation (Kostic et al. 2014).

12.4.1 Inflammatory Bowel Disease

Inflammatory bowel disease are a group of long-term inflammatory conditions that affect the digestive tract and alter the microbiota of the human GM. It causes inflammation of the GIT and affects the lining of the GIT. Examples of IBS disease include Crohn's disease, microscopic colitis, and ulcerative colitis (Kostic et al. 2014). The exact agent responsible for IBS is unknown, but environmental, genetic, microbial, and immunologic factors can contribute to IBS (Wehkamp et al. 2013). In the case of microbial shifts in IBD, some species are in abundance (*Escherichia coli*, *Fusobacterium*, *Serratia marcescens*, and *Candida tropicalis*), while others are reduced leading to alteration in metabolic pathways, increase in oxidative stress pathways, and decrease in basic metabolism with reduced SCFA production. These SCFAs, especially butyrate, are essential components of our nutrition and enable the immune system to produce more regulatory T cells that control excessive inflammatory responses. The lack of butyrate has been shown to lead to inflammatory bowel disease. Lipopolysaccharide is produced by the intestinal microbiota which is a source of acute inflammatory interactions that are able to activate innate inflammatory pathways, leading to the development of a chronic inflammatory state (Chiodini et al. 2015). Symptoms of IBS include persistent diarrhea, abdominal pains, bloody stool, rectal bleeding, weight loss, fatigue, etc.

12.4.2 Irritable Bowel Syndrome

Irritable bowel syndrome is a syndrome that affects the stomach and the large intestines. Microbial alteration is observed in IBS which records a reduction in *Actinobacteria* and *Bacteroidetes* and an increase in *Proteobacteria* and *Firmicutes*. Symptoms of IBS include abdominal bloating, abdominal pains, altered bowel habits, gas, constipation, etc.

12.4.3 Diabetes and Obesity

Dysbiosis can cause diabetes and obesity through altered energy regulation which leads to low variety of species, low production of SCFA, and slow use of molecules. SCFAs produced from the microbial fermentation of fiber play an important role in regulating blood sugar levels. SCFAs are absorbed in the blood circulation, where they affect glucose (sugar) storage in the muscle, liver, and fat cells. One specific type of SCFA, acetate, regulates appetite to decrease food consumption, which may reduce the risk of type 2 diabetes. A study carried out by some researchers comparing the community of gut microbes of lean and overweight twins in an experiment revealed that the gut community in lean people was diverse. *Bacteroidetes* broke down bulky plant starches and fibers into shorter molecules that the body can use as a source of energy (Tremaroli and Bäckhed 2012).

12.4.4 Cardiovascular Disease

Microbiome dysbiosis is implicated in CVD such as atherosclerosis, hypertension, and heart failure. Elevated levels of trimethylamine-N-oxide (TMAO) that is produced by certain species within the GM can increase the risk of heart disease. Such species may convert choline and L-carnitine—nutrients found in red meat and dairy—to TMAO, potentially increasing a person's risk of heart disease (Heidenreich et al. 2011). The human GM can improve heart health through promoting good cholesterol while moderating bad cholesterol.

12.4.5 *Clostridium difficile* Colitis/ Gastrointestinal Cancers

Clostridium difficile is a Gram-positive toxin- and spore-producing anaerobe. It is also one of the *Firmicutes* members in normal gut microbiota. The disease occurs when there is a disruption of normal healthy bacteria found in the colon due to medications for the treatment of infection. *Clostridium difficile causes disturbance of the normal flora of the colon which can result in colitis*. Due to the increase in cancer worldwide, several studies are being carried out to ascertain its course and possible treatment. Studies on the gut microbiome show that the gut microbiome has the ability to produce several metabolites and bioproducts that protect the individual and gut homeostasis. But microbial imbalance in the GM can allow the invasion of opportunistic microbes, leading to the production of high levels of toxins that can alter the GIT resulting in GI cancers. GI cancers occur when the host gut microbiome is impaired and the host immune system fails to restore homeostasis of the lining tissues. The individual's cell proliferation and death are impaired, immune functions are destabilized, and toxigenic microbes are in abundance in the GM (Song et al. 2020).

12.4.6 Celiac Disease

It's an autoimmune disease that contains certain genes and affects the immune system. Its occurrence is 1 in 100 predisposed individuals worldwide. Celiac disease damages the villi that line the small intestine when an individual consumes certain food that contains gluten. When celiac disease is left untreated it can lead to autoimmune disorders such as type 1 diabetes, dermatitis herpetiformis, and eventually bowel cancers (Shah et al. 2014). The gut microbiota that resides in the human colon plays a role in gluten metabolism. Microbes that are involved in the breakdown of gluten are *Lactobacilli and Bifidobacterium* spp. Symptoms of celiac disease are abdominal pains, chronic diarrhea, abdominal bloating, nausea, vomiting, etc. (Caminero et al. 2014).

12.5 Factors that Impact the Human Gut Microbiome

The factors that impact negatively on the GM include diet, pharmaceuticals, geographical locations, lifecycle stages, birthing process, infant feeding method, stress, etc. These factors have a negative effect on the health of GM. These negative factors can be worked upon to maintain a healthy GM.

12.5.1 Diet

The profile of microbial community within the human GM is affected by diet. Daily diet and nutritional status are the most important determinants of the human GM. The human balanced diet is made up of proteins, fat, and carbohydrates that supply vital nutrients, influencing the life span and physiologic state of many organs. The type and quantity of these diet constituents are greatly influenced by the community and diversity of the gut microbiome. The digestion of these constituents leads to variability in the end products, which play an important role in the development of the GM and the prevention, management, and treatment of certain diseases (Macfarlane and Macfarlane 2012). Diet has the ability to reshape the human GM and determine its composition. Daily diets can affect the GM balance by maintaining the beneficial microbes above the harmful ones, thus maintaining the health of the individual. When microbial balance is not maintained by diet intake the beneficial microbes are reduced and pathogenic microbes are more in the gut. This can lead to opportunistic microbes invading the gut leading to different diseases. Some dietary components that are not digested by host enzymes but remain unchanged through their passage to the gut may deliver a variety of growth-promoting and growth-inhibiting factors that influence the balance between microbial species, especially those responsible for the fermentation of different substrates within the microbial community. A normal and healthy human GM can induce alteration of the host physiology and affect the development of the immune system, neutralize carcinogenic compounds, synthesize vitamins, and play other roles. All these roles can be influenced by the type of diet consumed by the individual (Mansour et al. 2021). An unbalanced diet can lead to the progression and development of human diseases linked to the human GM. The diversity of microbes in the human GM can be restored and maintained by taking daily balanced meals with probiotics. Examples of strains used as probiotics include the following genera: *Lactobacillus*, *Bifidobacterium* and *Lactococcus*, *Enterococcus*, and *Streptococcus*. In the diet, fermented food, prebiotics, and probiotics should be included daily. More food with high resistant starch such as oats, green bananas, cooked potatoes, and legumes should be eaten. A high-fat diet can lead to gut microbiota dysbiosis, and this leads to increased gut permeability and metabolic endotoxemia (Mansour et al. 2021).

12.5.2 Pharmaceuticals

Medications especially antibiotics are necessary for the treatment of infections, but their misuse and underuse can affect the gut microbiome negatively. The chemicals present in the antibiotics can also affect the gut microbiome. Antibiotic intake alters the composition and magnitude of the gut microbiota, reduces species diversity, and eliminates antibiotic susceptible species thereby stimulating antibiotic resistance. Antibiotic-induced alterations in the GM cause persistent antibiotic-induced colitis which can reduce colonization resistance against the spore-forming, toxin-producing pathogen *Clostridium difficile*. The use of broad-spectrum antibiotics leads to overgrowth of *Clostridium difficile* which flourishes in the antibiotic-weakened gut, often resulting in a life-threatening condition (Theriot et al. 2014).

12.5.3 Geographical Locations

Individuals living in different parts of the world will have varying microbial populations based on different diets consumed by persons at that particular geographical location. Trillions of clusters of bacteria are found in the human GM which form an individual's bacterial ecosystem that dominates the gut. These clusters of bacteria are called the enterotype which enables researchers to ascertain the traits common among persons residing in various regions of the world. Enterotypes are classified into three namely *Bacteroides*, *Prevotella*, and *Ruminococcus*. Each person's enterotype is unique to the individual based on the fact that variations are observed due to several factors such as diet, life cycle stages, lifestyle, phytopharmaceuticals, infant feeding, etc. (Zoetendal et al. 2008; Bezirtzoglou and Stavropoulou 2011). The three categories of enterotype are affected by geographic location. *Bacteroides* are classified as enterotype 1 and they are made of microbes that derive their energy from proteins. People living in Western countries tend to have this enterotype because their long-term diet mostly consists of

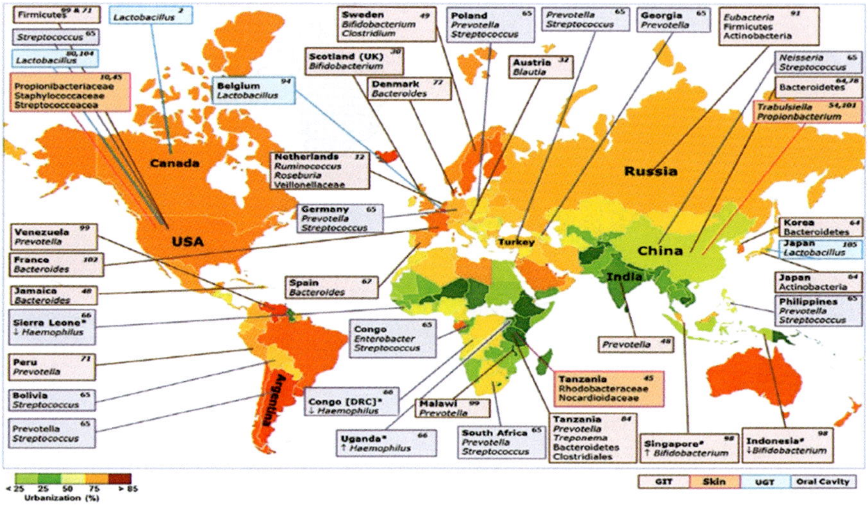

Fig. 12.1 Microbial populations across different geographical regions in the world. Source: Image credit: https://www.frontiersin.org/articles/10.3389/fmicb.2017.01162/full

fat and proteins. The *Prevotella* enterotype are classified as enterotype 2 and they are prevalent in non-Western countries where individuals eat a diet with high levels of fiber; these groups of microbes derive their energy from carbohydrates. These group of microbes derive their energy from carbohydrates such as sweets that are consumed by persons with high fiber. *Ruminococcus* (enterotype 3) are found basically in rural areas, and persons that have this enterotype consume a diet that is rich in resistant starches and dietary fiber (Konstantinidis et al. 2020; Stewart 2020). The different microbial populations associated with different regions of the world based on the impact of diet on each region are given in Fig. 12.1. This is in confirmation with the study carried out by De Filippo et al. (2010) which showed that the diversity of fecal microbes in children from rural Africa than that of developed communities in the EU. This suggests dietary differences in respect of the number of bacteria associated with the breakdown of fiber.

12.5.4 Lifecycle Stages

The stages of life can cause a variation in the human GM based on some factors and natural events of life. Such events include age, puberty, ovarian cycle, pregnancy, and menopause (Nicholson et al. 2012). From birth, an intimate relationship exists between the microbiota and the immune system; with aging there is reduced communication and the human GM develops imbalances in some species of microbes which are useful for maintaining health. The *Bacteroidetes* phylum bacteria tend to dominate numerically during youth, but numbers decline significantly by old age, whereas the reverse trend occurs for bacteria of the *Firmicutes* phylum.

12.5.5 Birthing Process

The birthing process, both cesarean and vaginal, plays an essential role in the development of the GM of an infant. Some studies have shown that the composition of the GM of an infant is directly linked to the delivery process. The microbiota of the infant is closely related to the microbiota of the mother from the first trimester. The vaginal pathway of delivery introduces bacteria that are enriched with vaginal secretions and fecal matter into the newborn, such as *Lactobacillus*, *Bifidobacterium*, *Escherichia coli*, and *Enterococcus* (Favier et al. 2002, 2003). *Bifidobacterium* and *Enterococcus* are involved in fermentation, which leads to the production of lactic acid that creates an acidic environment that defends the body against pathogens (Matsuki et al. 2003). The GM of a baby birthed by C-section varies because it does not pass through the birth canal where it comes in contact with microbes. The infant is exposed to bacteria from the skin of people, tools, and equipment used for the delivery. Bacteria associated with C-section delivery includes *Enterococcus*, *Streptococcus*, *Staphylococcus*, and *Propionibacterium* genera.

12.5.6 Lifestyle

Lifestyle factors that can impact the human GM are non-dietary factors such as smoking, excessive consumption of alcohol, stress, lack of exercise, etc. Smoking and lack of exercise have a negative impact on the human GM, by causing heart-related issues and dysbiosis in individuals. Stress is part of an individual's daily life, but chronic stress, depression, and anxiety can alter the microbes of the human GM. This is due to the physical changes in the microbiome and the structure of the intestinal lining. This leads to the secretion of stress hormones, inflammation, and autoimmune alteration that makes the gut release metabolites, toxins, and neurohormones that affect the mood of the individual and brain activity and alter eating behavior. Stress lowers the beneficial microbes in the GM for maintaining health that are responsible (Lutgendorff et al. 2008).

12.6 Maintaining a Healthy Human Gut Microbiome

A healthy human GM can be maintained with a well-balanced diet that is plant-based and high in fiber. It is also advisable to include lots of probiotic-containing foods in daily diets. Prebiotics which are non-digestible oligosaccharides that are commercially prepared can be added to diets or taken as pills to improve gut health. They aid in stimulating healthy microbes present in the gut. They are safe for consumption and not harmful to the health. A high-fat diet should be avoided because it can lead to microbiota dysbiosis which can affect metabolic activities in the GM. Food such as oats, legumes, cooked potatoes, beans, fruits, and natural yogurt are encouraged on a daily basis.

Lifestyle is important in maintaining a healthy GM. It is important that daily stress should be controlled due to its negative impact on the human GM. Smoking and excessive consumption of alcohol should be avoided. Water should be consumed to aid in bowel movement and prevent constipation. Medications should not be misused or underused to avoid the destruction of susceptible species and increase in drug resistance. Exercise is encouraged to prevent dysbiosis of the gut microbiome.

12.7 Conclusion

Microbes that reside in the gastrointestinal tract are made up of a dynamic community which revolves from birth to old age. The importance of the human GM in the overall health and development of individuals cannot be overemphasized. Its development is from the womb; thus women are advised to breastfeed their infants because breastmilk contains beneficial microbes that build up the GM of the infant and protect them from pathogenic microbes. The adult human GM is stable but can be influenced by diet, aging, environmental factors, and lifestyle behavior. Individuals are encouraged to maintain a healthy human GM by consuming a healthy balanced diet and practicing a healthy lifestyle.

References

Avershina E, Storro O, Oien T, Johnsen R, Pope P, Major K. Faecal microbiota shifts in composition and diversity with age in a geographically restricted cohort of mothers and their children. FEMS Microbiol Ecol. 2015;87:280–90.

Bengmark S. Ecological control of the gastrointestinal tract. The role of probiotic flora. Gut. 1998;42(1):2–7.

Bezirtzoglou E, Stavropoulou E. Immunology and probiotic impact of the newborn and young children intestinal microflora. Anaerobe. 2011;17:369–74.

Caminero A, Herran AR, Nistal E, Perez-Andres J, Vaquero L, Vivas S, et al. Diversity of the cultivable human gut microbiome involved in gluten metabolism: isolation of microorganisms with potential interest for coeliac disease. FEMS Microbiol Ecol. 2014;88:309–19.

Cheolmin S, Yong-Ku K. The complex interplay between gut-brain, gut-liver, and liver-brain axes. Chapter 3, the interactions between gut and brain in psychiatric and neurological disorders. Academic Press; 2021. p. 49–65. ISBN 9780128219270

Chiodini RJ, Dowd SE, Chamberlin WM, Galandiuk S, Davis B, Glassing A. Microbial population differentials between mucosal and submucosal intestinal tissues in advanced Crohn's disease of the ileum. PLoS One. 2015;10(7):e0134382.

De Filippo C, Cavalieri D, di Paola M, Ramazzotti M, Poullet JB, Massart S, Collini S, Pieraccini G, Lionetti P. Impact of diet in shaping gut microbiota revealed by a comparative study in children from Europe and rural Africa. Proc Natl Acad Sci U S A. 2010;107(33):14691–6.

Faith JJ, Guruge JL, Charbonneau M, Subramanian S, Seedorf H, Goodman AL, et al. The long-term stability of the human gut microbiota. Science. 2013;341:44–52.

Favier CF, De Vos WM, Akkermans AD. Development of bacterial and bifidobacterial communities in feces of newborn babies. Anaerobe. 2003;9:219–29.

Favier CF, Vaughan EE, De Vos WM, Akkermans AD. Molecular monitoring of succession of bacterial communities in human neonates. Appl Environ Microbiol. 2002;68:219–26.

Foster JA, McVey Neufeld KA. Gut-brain axis: how the microbiome influences anxiety and depression. Trends Neurosci. 2013;36(5):305–12.

Heidenreich PA, Trogdon JG, Khavjou OA, Butler J, Dracup K, Ezekowitz MD, Woo YJ. Forecasting the future of cardiovascular disease in the United States: a policy statement from the American Heart Association. Circulation. 2011;123(8):933–44.

Hollister EB, Riehle K, Luna RA, Weidler EM, Rubio-Gonzales M, Mistretta TA, Raza S, Doddapaneni HV, Metcalf GA, Muzny DM, et al. Structure and function of the healthy pre-adolescent pediatric gut microbiome. Microbiome. 2015;3:36.

Huurre A, Kalliomaki M, Rautava S, Rinne M, Salminen S, Isolauri E. Mode of delivery—effects on gut microbiota and humoral immunity. Neonatology. 2008;93:236–40.

Jiménez E, Marín ML, Martín R, Odriozola JM, Olivares M, Xaus J, Fernández L, Rodríguez JM. Is meconium from healthy newborns actually sterile? Res Microbiol. 2008;159(3):187–93.

Konstantinidis T, Tsigalou C, Karvelas A, Stavropoulou E, Voidarou C, Bezirtzoglou E. Effects of antibiotics upon the gut microbiome: (2020) a review of the literature. Biomedicine. 2020;8(11):502.

Kostic AD, Xavier RJ, Gevers D. The microbiome in inflammatory bowel disease: current status and the future ahead. Gastroenterology. 2014;146(6):1489–99.

Kovatcheva-Datchary P, Li Y, Xia Y, Xie H, Zhong H, Khan MT, Zhang J, Li J, Xiao L, Al-Aama J, Zhang D, Lee YS, Kotowska D, Colding C, Tremaroli V, et al. Dynamics and stabilization of the human gut. Cell Host Microbe. 2015;17(5):690–703.

Kulas T, Bursac D, Zegarac Z, Planinic-Rados G, Hrgovic Z. New views on cesarean section, its possible complications and long-term consequences for children's health. Med Arch. 2013;67:460–3.

Lin L, Zhang J. Role of intestinal microbiota and metabolites on gut homeostasis and human diseases. BMC Immunol. 2017;8(1):2.

Louis P, Flint HJ. Formation of propionate and butyrate by the human colonic microbiota. Environ Microbiol. 2017;19(1):29–41.

Lutgendorff F, Akkermans LMA, Soderholm JD. The role of microbiota and probiotics in stress-induced gastrointestinal damage. Curr Mol Med. 2008;8(4):282–98.

Macfarlane GT, Macfarlane S. Bacteria, colonic fermentation, and gastrointestinal health. J AOAC Int. 2012;95:50–60.

Mansour SR, Moustafa MAA, Saad BM, Hamed R, Moustafa RA. Impact of diet on human gut microbiome and disease risk. New Microb New Infect. 2021;41:2052–975.

Matsuki T, Watanake K, Tanaka R. Genus- and species-specific PCR primers for the detection and identification of Bifidobacteria. Curr Iss Intestinal Microbiol. 2003;4:61–9.

Morrison DJ, Preston T. Formation of short chain fatty acids by the gut microbiota and their impact on human metabolism. Gut Microbes. 2016;7:189–200.

Nicholson JK, Holmes E, Kinross J, Burcelin R, Gibson G, Jia W, Pettersson S. Host-gut microbiota metabolic interactions. Science. 2012;336:1262–7.

Qin J, Li R, Raes J, Arumugam M, Burgdorf KS, Manichanh C, Nielsen T, Pons N, Levenez F, Yamada T, et al. A human gut microbial gene catalogue established by metagenomic sequencing. Nature. 2010;464:59–65.

Rodriguez JM, et al. The composition of the gut microbiota throughout life, with an emphasis on early life. Microb Ecol Health Dis. 2015;26:26050.

Shah S, Akbari M, Vanga R. Patient perception of treatment burden is high in celiac disease compared to other common conditions. Am J Gastroenterol. 2014;109(9):1304–11.

Song M, Chan AT, Sun J. Influence of the gut microbiome, diet, and environment on risk of colorectal cancer. Gastroenterology. 2020;158(2):322–40.

Stewart L. What are microbiome Enterotypes and are they real? In: Atlas biomed blog | take control of your health with no-nonsense news on lifestyle, gut microbes and genetics.; 2020. https://atlasbiomed.com/blog/what-are-enterotypes/. Published 2020.

Theriot CM, Koenigsknecht MJ, Carlson PE Jr, Hatton GE, Nelson AM, Li B, Huffnagle GB, Li Z, J., and Young, V.B. Antibiotic-induced shifts in the mouse gut microbiome and metabolome increase susceptibility to *Clostridium difficile* infection. Nat Commun. 2014;5:3114.

Thursby E, Juge N. Introduction to the human gut microbiota. Biochem J. 2017;474(11):1823–36.

Tremaroli V, Bäckhed F. Functional interactions between the gut microbiota and host metabolism. Nature. 2012;489(7415):242–9.

Wehkamp J, Antoni L, Ostaff M, Stange EF. The intestinal barrier in health and chronic inflammation. In: *Current understanding and implications for future therapeutic intervention*. Falk Foundation eV; 2013.

Zoetendal EG, Rajilic-Stojanovic M, de Vos WM. High-throughput diversity and functionality analysis of the gastrointestinal tract microbiota. Gut. 2008;57:1605–15.

13 Chemical Safety and Chemical Security: A Guide to Preventing Health Hazards

Kaine Okorosaye-Orubite

13.1 Introduction

13.1.1 Definition of Terms

13.1.1.1 Chemicals

Chemicals are substances that have a definite composition, and they can be natural, for example water, or artificial, such as sodium hydroxide. Natural chemicals are found in nature like in plants and animals, while artificial or manufactured chemicals are produced by scientists. There is actually a minimal difference between natural and artificial chemicals of a particular compound. If a chemical found in nature is synthesized in the laboratory, there is no distinctive difference between them, for example, vitamin C in oranges is the same as the manufactured one.

Some chemicals are used directly or as precursors to the manufacturing of other chemicals. For instance, phosphorus compounds are precursors for manufacturing agrochemicals and plastic additives among others. We come across artificial or manufactured chemicals in the home such as bleaches, dishwasher powders, turpentine, swimming pool cleaning liquids, weed killers, pest control products and the likes. It is important that precautions are taken while using chemicals at home.

13.1.2 Health Hazards

In the context of this work, health hazards are limited to chemical health hazards. The term 'chemical health hazard' refers to the properties of a chemical that can cause acute or chronic health problems. A chemical health hazard occurs when one is exposed to chemicals that have the potential to cause harm to health and life. Chemicals can be inhaled, ingested, swallowed, or even absorbed through the skin (Nwaichi, 2018). Acute health issues are such as burns, vomiting asphyxiation, etc., and chronic problems are cancer, liver damage and so on.

13.1.3 Chemical Safety

Chemical safety involves the protection of human health and the environment (Nwaichi, 2018) by controlling the extent of exposure to potentially hazardous chemicals. Simply put, chemical safety is protection against accidents while handling chemicals whether in the laboratory or our homes. It covers both natural and manufactured chemicals. Chemical safety activities range from

K. Okorosaye-Orubite (✉)
Department of Pure and Applied Chemistry, University of Port Harcourt, Port Harcourt, Nigeria
e-mail: orubite@uniport.edu.ng

E. O. Nwaichi (ed.), *Modernity in Health and Disease Diagnosis: The Account from STEM Women*, Sustainable Development Goals Series, https://doi.org/10.1007/978-3-031-34963-8_13

exposure situations of the chemicals in the environments to synthesis, production, transportation, use and disposal. These are all summed up in the topic 'chemical management'. The WHO in its 59th world health assembly issued resolutions and decisions (WHA 59.15, strategic 'Approach to international chemicals management', 2006) to ensure chemical safety and hence minimize hazards and accidents. Implementing a good chemical safety policy in an organization will not only save lives but also protect the environment. Laid down procedures to ensure the safety of personnel, the community and the environment are pertinent even in our homes.

13.1.4 Chemical Security

Chemical security centers on preventive measures designed to reduce the risk of deliberate removal (theft) and misuse of chemicals with the intent of committing a crime or causing harm. According to OPCW (Organization for the Prohibition of Chemical Weapons), chemical security refers to measures used to prevent deliberate releases of toxic chemicals and to mitigate the impact if such events occur. In a wider context, it also includes policies to prevent attempts to acquire toxic chemicals or chemical weapons precursors.

In essence, chemical security is concerned with:

- Prevention of terrorist attacks on industrial plants.
- Theft of chemicals with potential hazards.
- Contamination of chemicals or products and degradation (CCPS 2008).
  Chemical security comprises:
- Methods or processes of identification of possible hazard.
- Risk assessment designs in handling chemicals.

A risk assessment according to Kaplan and Garrick (1981) should answer such questions as:

- What can go wrong?
- How likely is it?
- What are the consequences?

13.2 Relationship Between Chemical Safety and Chemical Security

Both chemical safety and chemical security are geared towards protecting life and the environment. They are all necessary in the research setting as well as in our homes.

13.2.1 Chemical Safety Centers on

- Preventive measures in the design and control of a chemical.
- Identifying vulnerabilities.
- Likelihood of accident and control.

For example, in the laboratory, corrosive or hazardous chemicals are labelled and kept separately or signs like 'dispose properly could cause damage to plants' placed on them.

Summarily, chemical safety depicts the likelihood of accidents, consequences and control while using chemicals.

On the other hand, chemical security is the degree of protection against the misuse of a chemical in terms of the damage it could cause to life and the environment through theft, loss or criminal activities.

Chemical security measures in a facility may include:

- Physically protecting employees (providing security services for employees).
- Safeguarding unauthorized access to get chemicals or preventing deliberate attacks (building fences, security posts, etc.)
- Safeguarding against the release of hazardous chemicals to the environment.

Broadly speaking, chemical safety aims at precautions taken during the intentional use of a chemical. Chemical security further highlights possible unintentional use or misuse of the chemical that can be of danger to life and the environment while soliciting measures to guard against such activities. As an analogy, a researcher working with ammonium chloride wears personal protective clothing (PPE) and observes all laboratory rules (intentional use) in chemical safety. Safeguarding ammonium sulphate from getting into the wrong hands (unintentional use) is chemical security.

A proactive approach to effective chemical safety may include such steps as:

(a) Identification of chemicals being handled or stored. All chemicals available should be labelled properly; identified and separately stored according to their hazard levels; likewise, risk assessment of each chemical should be recorded. For example, in a laboratory, inflammable chemicals should be stored away from others with labels clearly showing so. In the home, used bottles of other chemicals should be re-labelled.
(b) Hazard prevention and controls.
A record of the hazard levels of each chemical should be explicitly stated with possible control measures. This will give information to anyone in case of an emergency. Knowledge about chemical hazards should be shared with all involved.
(c) Workers in any organization should be trained regularly on how to handle chemicals. Training and educating staff are critical to maintaining safety.
(d) Safety efforts should be reviewed periodically to access the effectiveness of safety programs.

13.2.2 Conflict Between Chemical Safety and Chemical Security

Information on chemical safety can constitute a chemical security risk. In the laboratory, chemical safety measures require that every chemical should be labelled and separately stored to identify hazardous ones. Everyone should have knowledge of this information. Labels identifying these chemicals can also be targets for theft risk and misuse. Aqua Regia, a concentrated chemical of three concentrated acids, is often labelled and separately stored. An employee or even thieves can get access to this chemical for the wrong purposes. Displayed labels easily attract attention and such chemicals are susceptible to theft or misuse. Those who attacked people with acid baths identified the acid through the labels.

Sharing knowledge about a chemical can expose other uses of the chemical which can inspire security risk.

13.2.3 Resolution of Conflict Between Chemical Safety and Chemical Security

The conflict between chemical safety and chemical security can be resolved by observing chemical management best practices.

A chemical management system ensures:

- Proper compliance with safety, environmental and food safety policies.
- Chemical procurement and purchasing.
- Safe storage, handling and use of chemicals.
- Disseminating information about hazardous chemicals to the community in compliance with laid down regulations, having MSDS (Material Safety Data Sheet).
- Preventing exposure of workers and the environment to chemicals.
- Proper waste disposal.

13.3 Why Chemical Safety and Chemical Security Awareness?

Chemical safety and chemical security awareness have become necessary because of recent happenings around us. Every day we are exposed to thousands of artificial chemicals; a report by the European Environmental Agency (2020) attributes estimates of about 6% of the world's disease burden and 8% of deaths to chemicals. Unauthorized and misuse of chemicals by criminals have resulted in several casualties: the bombing of factories, schools, oil installations or pipelines; fire outbreaks; insurgency and militancy; acid baths; deliberate food poisoning; etc., were traced to the use of chemicals. Also, explosives and Improvised Explosive Devices (IEDs) made from chemicals have found to be used in these attacks. As a result, various countries have promulgated laws to protect the use of chemicals to forestall these criminal acts. The Nigerian Senate passed the 'Chemical weapons prohibition Bill' on Wednesday, 21 May 2018, and placed some chemicals on the prohibition list.

Some of them are:

- Ammonia sulphate.
- Potassium sulphate.
- Urea.
- Hydrogen peroxide.
- Hexamine.
- Urea-formaldehyde.
- Sulphur powder.
- Ammonia powder.
- Nitroglycerine.
- Sodium chlorate.
- Liquid ammonia.

(Source: http://www.opcw.org/index.html)

Many scholars have observed and reported deficiencies in the observation of chemical safety and chemical security in our environment. Research institutions and homes are places where ignorance and negligence have caused a lot of chemical health hazards (Youssef, 2018). Educational and research institutions are slower in implementing safety and security programs than the industries; this situation has been reported in many forums. Walters et al. (2017) observed deficiencies in Trinidad's institutions, particularly in the areas of hazard identification and emergency response. The prevalence of chemical hazards among laboratory workers in some public universities in Lebanon was assessed by Nasrallah (2022). Regular training on laboratory safety measures is essential to ensure health and safety. In some laboratories, chemicals are abandoned and some are not properly stored. Laboratories are supposed to have adequate light and running water, and in some of our tertiary institutions this is not the case. There is an unsatisfactory level of chemical safety and chemical security awareness in school laboratories. There has to be more awareness, change in attitude, and standard practices among tertiary and secondary students on chemical safety in the laboratory. Employees are often exposed to chemical compounds on numerous occupational settings, and yet there are significant gaps in knowledge and scare preparedness in adherence to safety procedures. Employers and heads of institutions must enforce measures to ensure chemical safety and chemical security. Simple laboratory guidelines will be given in this write-up.

Chemicals are used every day in our homes—cleaning fluids, paints, glues, pesticides and even medicines. Lack of awareness of chemical safety in homes also resulted in many avoidable domestic accidents. In the home environment chemical safety and chemical security awareness also needs some emphasis to ensure healthy homes. With thousands of chemicals in the local market and many more introduced to us each day, awareness is obligatory. Farmers and their families who are constantly in contact with farm chemicals are apparently ignorantly suffering from acute and chronic effects of small quantities of chemicals they come across daily. Are you observing necessary precautions at home? Do you know that all cleaning agents are chemicals and are harmful? What about spraying pesticides to kill that cockroach or any vermin? Prevention is better than cure.

13.4 Some Common Use Chemicals and their Health Hazards

Chemical health hazards may be acute in which case the effect is felt within hours or even days of occurrence or chronic when it takes a while to manifest symptoms of an illness. Sources of contact with these chemicals may be through inhalation, swallowing or absorption through the skin. Table 13.1 shows examples of common chemicals and possible hazards they can cause while in use.

Table 13.1 Chemicals and possible health hazards

S/N	Chemical	Place used	Likely hazard
1	Acids	Laboratory	Burns, death on ingestion
2.	Ammonium sulphate	Food companies, fertilizers, air conditioners, fire extinguishers, etc.	Long-time use can cause acidification of soil, skin irritation and affect breathing when inhaled
3.	Sulphur powder	For skin irritation, acne and sumac infections	Causes diarrhoea when ingested
4.	Isocyanides	Used as paint additives	Destroys the immune system
5.	Chloroform	A precursor to many refrigerants	Tumours in the liver and kidney
6.	Methanol	Manufacture of pharmaceutical products	Blindness and lethal even in small amounts
7.	Sniper (pesticide)	Used locally to kill cockroaches and other insects	Highly poisonous when ingested. Kills even animals that eat the dead insects
8.	Hypo (bleach solution)	Used as a cleaning agent or for bleaching white fabric	

13.5 Who Is Vulnerable to Chemical Health Hazards?

Everyone is vulnerable to chemical health hazards and attacks. Vulnerability simply refers to a state of being weak or helpless in a situation. Someone who is vulnerable is weak and without protection, as a result, they can be easily hurt physically or emotionally.

Hazardous chemicals are present in the air and in consumables both at the workplace and the home. According to the WHO Director-General, during the Ministerial Dialogue held on 7 July 2021, at the Berlin Forum on Chemicals and Sustainability: Ambition and Action Towards 2030, deaths due to hazardous chemicals worldwide rose from 1.56 million in 2016 to two million in 2019. The report further highlighted that the world experiences daily deaths of between 4270 and 5400 due to unintentional exposure to chemicals.

One major source of death due to hazardous chemicals is lead poisoning. Lead is used to enhance the quality of some paints. The WHO Director-General also reported that most of the deaths especially unintentional poisoning are among children and young adults who are more vulnerable (WHO webinar 2022).

Employees at chemical factories and laboratories are vulnerable to a lot of hazards as they are exposed to chemicals for longer times. Such chemicals include detergents, paints, cleaning agents, etc (OPCW conference 2019).

The home setting is vulnerable too. A simple illustration of vulnerability at home is keeping alcohol (spirit) where there are underaged children. The likely hazard here is that the children may get access to it, get drunk and be intoxicated. Alcohol or alcoholic beverages should be stored out of reach of children who are vulnerable to the hazard of its dangers. A child who is always sent to buy a cigarette for the parent(s) is vulnerable to the use of cigarettes or other drugs (SDG 3).

13.6 Vulnerability Assessment

It is important to recognize vulnerabilities and do whatever is necessary to reduce or eliminate risks.

13.6.1 What Are the Areas of Vulnerability?

13.6.1.1 In our Homes, Vulnerabilities May Include

1. Purchase of chemicals from illegal sources. Cleaning agents, fuels and medicines are common chemicals used in the home. Most of these products are not registered by relevant control agencies. Purchasing from such avenues leaves the home vulnerable to chemical health hazards. Locally produced disinfec-

tants and cleaning solutions can be sources of vulnerability to chemical health hazards in the home. Due to poor economic conditions, some people are prone to patronize these products. They really pose threats to health.

2. Improper information about the other uses of the chemical.
   Some chemicals have dual uses. They would have bought it for a particular purpose, but other uses of the chemical can make it a chemical security risk.
3. Use of adulterated chemicals.
   Many adulterated chemicals are all over the market. Several cases have been reported where kerosene explosions occurred due to adulterated products. Kerosene is the main source of cooking and lighting especially for the majority of poor citizens as an alternative to electricity and gas. In Southern Nigeria, the oil-producing area of Nigeria, illegal refineries (clamped down by the government of the country) have contributed to a lot of kerosene explosions (Olugbenga 2005). Many unregistered pharmacies also proliferate our environment; medicines should be purchased in only reputable drug stores. The rural communities in Nigeria are vulnerable to patronizing quakes as drug vendors. The local authorities (Local Government Health office) can minimize or eliminate this risk. Premium Motor Spirit (PMS) or petrol and diesel are products often adulterated which presents risks. Apart from vehicles, both products are also used in electricity generators.
4. Carelessness on the part of users to storage and disposal.
   Carelessly storing chemicals can be a vulnerability risk in the home. Sometimes chemicals are transferred to other containers that bear different labels. In one of our discussion forums, a lady reported how she stored a cleaning solution in an empty Sprite bottle and her child mistook it for water and drank it. An early intervention saved the situation. Another child drank kerosene kept in the kitchen at the reach of a toddler who also ingested the substance. Serious health emergencies are often recorded when chemicals are carelessly kept in the home.
5. Disposing of used or unwanted chemicals also poses vulnerability risk. I had an experience in which I used pesticides to kill cockroaches and disposed of the dead ones just on the field in my backyard. A few days later I observed all the free-range chickens I keep were either dead or sick. From investigation, I realized they consumed the cockroaches killed by the pesticide. That was a careless disposal resulting in a chemical health hazard.
6. Other family members are not aware of chemicals in the home.
   When other family members are not aware of a new chemical in the home, it can be a vulnerability risk. A woman who produces 'Hypo' (the name for locally made bleaching solution in Nigeria) used a bucket that looks like the bucket for storing drinking water in their home. Her daughter dipped her cup into the bucket to quench her thirst. If Mummy had informed the family members, that casualty would have been averted.

The following guides are pertinent to forestall vulnerability risk:

Taking inventory of all the household chemicals used in the home and their proper storage systems. This calls for reading the label on containers of all chemicals purchased. Everyone should form this habit so as to avoid possible hazards. Some chemical products are best kept in the refrigerator, some in the freezer, and some on the shelf or cupboard. Medicines are also chemicals; vulnerability assessment includes keeping them out of the reach of unauthorized persons. Most chemical products will have directions for use on the container as well as dangers in applications. Identify the likely hazard and risks involved in using any chemical product.

One should also be informed about first aid activities to apply in case of emergency and assess how likely such hazard can happen and the consequences.

13.6.1.2 In a Research or Laboratory Facility

1. Chemical logos are not properly placed on chemicals. Labels are used to identify chemicals, but once they are in a container it is difficult to place that chemical. It is therefore of utmost importance that such labels be properly fixed on the container. Imagine a scenario where you have distilled water and a dilute acid, both in an unlabelled container, there is the likelihood of mistaking one for the other.
2. Transporting chemicals through students, commercial passengers or personal vehicles.
   Often times we take for granted this area of vulnerability assessment. One should be very mindful of how chemicals are transported. We find in some of our laboratories how students are allowed to purchase their own chemicals for research work when actually chemicals should be issued to them by laboratory staff. The usual response is 'we ran out of stock'. From the part of the world where I come from this is almost a norm but seriously against laboratory best practices. Students may not have the full information on how to handle such chemicals while transporting them to the laboratory. Again misuse, diversion and even accidental spilling or other forms of mishap can occur. Some even use commercial passenger vehicles during this process.
3. Security operatives are not trained on chemical safety/security to enforce chemical security measures.
   A security agent who is not trained to recognize chemicals and their hazard levels may not be aware when that chemical is stolen from a facility he protects. It means therefore that the lack of training on chemical security and general science education (Nwaichi et al. 2021) makes security operatives and other workers, even the facility itself, vulnerable to chemical health hazards. It will be interesting to investigate how many security personnel in our tertiary institutions have such awareness and education. This could be a time bomb ticking.
4. Inadequate protection of the laboratory and facilities against theft and loss.
   We should make sure laboratories are protected against both inside and outside theft. An inside theft is when chemicals are removed by workers who have access to them and outside theft is from an external source. Vulnerability assessment procedures must be put in place to militate against theft and loss of lives (SDG 3).
5. Working with a disgruntled staff that could steal or sell chemicals for a pittance.
   Even though human beings are insatiable, an underpaid staff like we find in some research laboratories can hardly be controlled. A hungry man is an angry man; where there are security lapses, a disgruntled employee will be ready to sell chemicals not minding the after-effect of their actions. Therefore, employers in these facilities must consciously guide against dissatisfaction among staff (SDG 3).
6. Purchase of chemicals from illegal distributors.
   This is also a vulnerability risk that should be checked. Illegal distributors are vanguards of fake and substandard chemicals. Their products may be cheap but not achieve the purpose of purchase. Some may even cause death.
7. Insufficient information on chemicals used.
   Not having sufficient information about the chemical used is a vulnerability risk. For instance, petrol is a chemical that is inflammable near any flame or fire, and a researcher who is ignorant of this is at risk of a fire accident. Some chemicals are to be kept in dark places not to be exposed to sunlight and so on.

13.7 World Concern for Chemical Safety and Chemical Security

In order to live in a world free of chemical weapons and safe usage, the Organisation for the Prohibition of Chemical Weapons (OPCW) was created on 29 April 1997 with 193 member states. The goal is 'to prevent chemistry from ever again being used for warfare thereby strengthening international security'. OPCW cooperates with the chemical industry to educate and reach out to society about chemical safety. To this end in Nigeria, the United States Department of State's Chemical Security Program (CSP) in conjunction with associations like the Chemical Society of Nigeria (CSN) holds workshops and training for researchers, industrialists, and manufacturers to create awareness and sensitization in chemical safety and chemical security (Nigerian Regional ChemicaSecurity Workshop 2018; Chemical Security and Vulnerability Assessment and Security Plan Development Workshop 2020).

13.7.1 How do Chemicals Get to Society?

(a) Importation.
Chemical importation is controlled by various countries under their regulatory bodies. In Nigeria, this is achieved through proper and strict monitoring by the regulating body National Food and Drug Administration Control (NAFDAC). Their control covers:
  - Issuing permit for the importation of all categories of chemicals.
  - Listing of all chemical importers and marketers.
  - Inspection of storage facilities when chemicals are imported.

(b) Illegal local manufacturers.
Illegal manufacturers still find their products into society beating all regulatory controls. The unsuspecting consumer suffers the effects of any substandard product. In Nigeria, there are many local manufacturers of household chemicals who are not NAFDAC approved even though the law forbids them. All chemicals sold should have a NAFDAC registration number. But how many do have them? The ignorant society patronizes them.

(c) Industries.
Most chemical industries are licensed to import their raw material. They become avenues where chemicals get into society. Fraudulent practices still bring some of these chemicals into the open market.

(d) Universities and research institutes.
Universities and research institutes are legally permitted to import chemicals. Laboratory chemicals are consumables in these places. Sometimes these chemicals may find their way into the wrong hands in society.

(e) Theft and insincerity among authorized users.
Theft of chemicals can occur at the point of importation without the knowledge of the legal importers, or it could be theft in their storage facility. Chemicals can also be brought into society through authorized end users. Sincerity among employees is another avenue where hazardous chemicals get into the society.

(f) Diversion of chemicals.
Sometimes genuinely imported chemicals are diverted to other sources. Despite efforts by regulatory bodies, industrial and laboratory chemicals are being diverted to purposes for which approval was not obtained.

In summary, chemicals for safety and security concerns may be imported for a genuine purpose by authorized importers, industries, and research institutes. Insincerity among authorized users can divert such chemicals to society. Illegal local manufacturers are also sources of distribution of chemicals of concern. Some of them are dual-use chemicals that are even in homes.

13.7.2 Dual-Use Chemicals

Dual-use chemicals are chemicals that can be used for both beneficial and harmful purposes (Douglas et al. 2014). Examples are:

(a) Dimethyl methyl phosphate (DMMP) is a flame retardant used for furniture, buildings, upholstery making, and in the electrical industry. This chemical is a chemical agent nerve precursor.
(b) Pesticides/rodenticides. These are used to get rid of pests and rodents in our environment and homes. They can be poisons to us too.
(c) Hydrogen peroxide, (H_2O_2), finds use in the recipe for sanitizers. Sanitizers are important chemicals in 'the new normal' in our world today. Do you know that (H_2O_2) is a dual-use chemical used in the manufacture of explosives?

In our homes, we come across common dual-use chemicals and chemical products such as:

(a) Hypo (bleach solution).
(b) Insecticides, pesticides, and rodenticides.
(c) Sniper.
(d) Ammonium nitrate (fertilizer).
(e) Camphor balls.

All these are beneficial to us in the home but at the same time, they can be harmful. Hence proper management and handling is very important (www.csp-state.net).

13.8 Guides to Preventing Health Hazards

Some useful guides to help reduce/eliminate chemical health hazards are as follows:

(a) In the Laboratory.
1. Purchase chemicals from legal and approved distributors.
2. Ensure proper storage of chemicals.
3. Keep chemicals away from the wrong persons.
4. In the laboratory regularly keep charts of chemicals dispensed so as to identify loss.
5. Organize chemical safety and chemical security awareness training for staff, students, and security personnel in tertiary institutions.
6. Always use Personal Protection Equipment (PPE).
7. Be aware of the dual-use properties of chemicals.
8. Properly dispose of any chemical waste.

(b) At Home.
1. Always label home-used chemicals properly. Salt and granulated sugar have the same appearance, differentiate them by labelling their containers.
2. Make sure rodenticides and pesticides are not kept near food or food materials. When rodents or insects are killed make proper disposal of them by digging a hole; they could be poison to other animals in the environment if carelessly disposed of.
3. Let everyone in the home know about the dangers of dual-use chemicals available in the home.
4. Do not take chemicals (hazardous or not) out of their original containers. If you must, properly label the container or inform family members.
5. Always wash your hands or parts of your body with soap and water after using chemicals.
6. When you buy chemicals, read the labels, for there are usually directions for use and warnings about their hazardous nature.
7. Keep chemicals out of the reach of children. For older children explain the dangers of hazardous chemicals to them.
8. Keep chemicals in original containers, if you must switch containers, please label them properly.

13.9 Some Useful First Aid Tips to Apply in Chemical Hazard Emergencies

The tips or information given in Table 13.2 are intended to serve as first aid before visiting a medical facility.

Table 13.2 First aid in chemical hazard emergencies

S/N	Type of emergency	First aid treatment
1.	Skin or eye contact	Flush with water copiously until irritation subsides. Do not rub, wipe, or apply any medication to the affected area.
2.	Burns	Wash off the chemical with water until it is completely removed. You can cover with a clean bandage or gauze. Do not apply pressure on the burned skin. Visit a medical facility immediately for further treatment.
3.	Inhalation	Move out from the area where the fumes are and get fresh air immediately. If the person is unconscious, those around should provide cardiopulmonary resuscitation (CPR).
4	Ingestion	Make the person expel what is in the mouth as much as possible. Give milk or water before medical help arrives. If it is at home, some traditional old method advice is that the person be given palm oil which aids the person expel any remnant inside.
5	Poisoning	Lime juice or vinegar can be used to neutralize ingested poison. Palm oil is recommended in this case too.
6.	Acid ingestion	The victim should be given milk or water if conscious enough before medical help. Note the type of acid, its concentration, and the volume probable swallowed. Acid can be neutralized in the mouth by rinsing with water and fluoride paste.
7.	Other solutions	Identify the type of solution and use large volumes of water to rinse affected parts.

13.10 Conclusion

There is poor adherence to chemical safety and chemical security best practices in laboratories as well as our homes. Health is wealth, so many health hazards can be avoided by observing chemical safety and chemical security practices. In a technology-laden world, everyone comes across chemicals whether one is a scientist or not, for sure we all live in homes. Have we bothered about the security of the chemicals used? How many persons have access to them for other purposes apart from the intended use? Since chemicals are used in one form or the other by everyone, we are all vulnerable and must strive to maintain a safe environment and live healthy lives. We can be our brother's keeper by reporting illegal manufacturers to the relevant controlling agencies in our countries. The precautionary saying is 'better safe than sorry'.

References

Aspects of chemical security: Dual –use chemicals and international controls. From: www.csp-state.net. Accessed April 12th 2021.

CCPS. Guidelines for Chemical Transportation Safety, Security and Risk Management. 2008. Retrieved from https://www.aiche.org/resources/publications/books/guidelines-chemical-transportation-safety-security-and-risk-management-2nd edition from Center for Chemical Process Safety.

Chemical security and vulnerability Assessment and security plan development workshop. 11–14 February 2020, Port Harcourt, Nigeria.

Douglas BW, Ho P, Hardesty J. Safety. Security and dual -use chemicals, Journal of Chemical Health, and Safety. 2015;22(5):3–16.

Kaplan S, Garrick BJ. On the quantitative definition of risk. Risk Anal. 1981;1(1):11–27.

Nasrallah IM, El Kak AK, Ismaiil LA, Nasr RR, Bawab WT. Prevalence of accident occurrence among scientific Laboratory Workers of the Public University in Lebanon and the impact of safety measures. Saf Health Work. 2022;13(2):155–62. https://doi.org/10.1016/j.shaw.2022.02.001.

Nigerian Regional Chemical Security Awareness Workshop, 7–9 March 2018. Organised by the chemical Society of Nigeria and sponsored by the united State's Department of State, Port Harcourt, Nigeria.

Nwaichi EO. Responsible conduct of Chemical Sciences Research and challenges in Nigeria and West Africa. In: Responsible Conduct and Ethical Practice in Chemical Sciences Research, Safety, Security, Education, and Risk Management: The Catalytic Role of the Professional/Learned Society. American Chemical Society (ACS) Books. ACS Symposium Series, Vol. 1288. Chapter 16, 2018. pp. 223–237.

Nwaichi EO, Ugwoha EO, Ogu RN. Science education: a veritable tool for development. In: Science by women:

stories from careers in STEM. Springer Nature; 2021. p. 5–29.

Olugbenga SA. Adulterated kerosene burn disaster: the Nigeria experience. Ann Burns Fire Disaster. 2005;18(1):40.

OPCW conference of the States parties (2019),-Engaging the Chemical Industry Associations. Note by the Director General C24/DG.17at the24th session 25–29 November 2019: pp. 1–12.

Walter AUC, Lawrence W, Jalsa NK. Chemical laboratory safety awareness, attitudes and practices of tertiary students. Saf Sci. 2017;96:161–71.

WHO. Webinar on Human Health Risk Assessment – Updated WHO toolkit. 25th January 2022. Departmental news Accessed 25th August 28, 2022.

Youssef A. Chemical safety and security challenges in academic institutions in developing countries; 2018. https://doi.org/10.1021/bk-2018-1288.ch006.

Chemical Leaching into Food and the Environment Poses Health Hazards

14

Gloria Ukalina Obuzor
and Uchechi Bliss Onyedikachi

14.1 Introduction

The quest for urbanisation and economic advancement globally has resulted in a more significant proportion of people living in townships; hence, more plastic substances are littered indiscriminately into the environment daily, leading to plastic pollution (Mngomezulu et al. 2020). More than 8 billion tonnes of plastic have been manufactured by humanity since 1950; more than half of this plastic ended up in landfills, and only approximately 9% was recycled (World Population Review 2022). Nigeria is the highest producer of plastic waste in Africa (about 18,640 tonnes) and is ranked tenth globally (World Population Review 2022). A section of plastic dump in a landfill located in Aba, Abia State, Nigeria (Fig. 14.1), indicates that landfills and aerial deposition are a major nexus for plastic leachates dispersal into the ecosystem. Plastic may harm the ecosystem by releasing poisonous compounds into the soil and groundwater that suffocates or kill animals who unintentionally swallow it. Finding waste management strategies and technologies that are environmentally friendly, economical, and socially feasible has been a continual desire for both leaders and technocrats as the manufacturing of plastic garbage has become unavoidable (Cohen 2018). The majority of our food, drink, packages, toys, clothing, sporting goods, and electrical components are manufactured or packaged using plastic components, which poses a serious health risk if inhaled or consumed. Plastic leaching and build-up of component monomers, endogenous additives, and absorbed ambient contaminants are all potential causes of chemical toxicity. The majority of tools used at home and work are made of plastic. Organic monomers are used to create plastics partly or entirely, making them strong, lightweight, and durable to create plastics (Mohanty et al. 2022). Cellulose, coal, natural gas, salt, and crude oil are some natural materials that may be processed by a polymerisation or polycondensation process and can be moulded or shaped in various ways according to their intended use (Jem and Tan 2020). Numerous varieties of plastic are produced using a range of chemicals. For instance, only a container made of plastic might contain more than 4000 distinct compounds. Chronic leachate exposure is considered very hazardous due to the potential accumulation and health impart over time (Wright and Kelly 2017). Their migration in the soil is also affected by physical processes such as volatilisation, adsorption, and dissolution in soil pore

G. U. Obuzor (✉)
Department of Pure and Applied Chemistry, University of Port Harcourt, Port Harcourt, Nigeria

U. B. Onyedikachi
Department of Biochemistry, College of Natural Sciences, Michael Okpara University of Agriculture, Umudike, Abia State, Nigeria
e-mail: ub.onyedikachi@mouau.edu.ng

E. O. Nwaichi (ed.), *Modernity in Health and Disease Diagnosis: The Account from STEM Women*, Sustainable Development Goals Series, https://doi.org/10.1007/978-3-031-34963-8_14

Fig. 14.1 A cross section of a plastic-polluted area in Aba, Abia State, Nigeria

water, and these alongside the time of human intervention determine the extent of impact in the soil (Ahmed et al. 2022). Most places, however, lack proper waste management, and the plastic pollution risk is anticipated to be exposure levels dependent, although there is a dearth of cogent evidence about the exposure levels (Chang et al. 2022). The local population is at severe health risks such as hormone-related cancers, infertility, and neurodevelopment disorders like ADHD and autism. Research has also confirmed that plastic wastes are carriers of pathogenic bacteria and viruses, further hampering the spread of diseases (Niyobuhungiro and Schenck 2021; Meng et al. 2021). The classification of plastics, sources, mode of exposure, effects of additives, the health implications of plastic leachates to aquatic and soil organisms and humans, remedies to the effect of plastic leachate pollution, and alternatives to plastic are emphasised in this chapter.

14.2 Plastics Are Classified According to Physical and Chemical Properties

14.2.1 Polyethylene Terephthalate (PET or PETE)

One of the most commonly used plastics is PET. It is robust, translucent, lightweight, and frequently used in textiles and food packaging (polyester). Examples include beverage and food bottles (for salad dressing, peanut butter, honey, and others) and polyester clothes. The properties include a moisture barrier, toughness, transparency, solvent resistance, and gas softening at 80 °C. It results in cancer, nausea, and diarrhoea (Mikhailovich and Fitzgerald 2014).

14.2.2 High-Density Polyethylene (HDPE)

The world's most widely used plastic is polyethene, divided into three types: linear low-density, low-density, and high-density. Due to its solid nature and resistance to moisture and chemicals, HDPE is perfect for use in pipelines, cartons, and other building supplies. Examples include rigid pipes, toys, buckets, park chairs, detergent bottles, cereal box liners, and milk cartons. It has a waxy surface, is opaque, softens at 75 °C, is hard to semi-flexible, is resistant to chemicals and moisture, and can cause stomach ulcers (Eriksen et al. 2019).

14.2.3 Polyvinyl Chloride (PVC or Vinyl)

Polyvinyl chloride can resist chemicals and weathering, making it suitable and highly desirable for building and construction applications. The fact that it is a poor conductor of electricity, tough and rigid, makes it popularly used for high-tech like lines and cables. It is also used in healthcare systems because it is impermeable to germs, simple to clean, and offers single-use applications that prevent infections. However, PVC, commonly employed in medical applications, is the plastic that poses the greatest threat to human health as it is known to release harmful leachates (e.g., lead, dioxins, vinyl chloride). Plumbing pipes, credit cards, toys for people and animals, rain gutters, teething rings, IV fluid bags, medical tubing, and oxygen masks are a few examples. Properties include durability, robustness, softening at 80 °C, transparency, weldability plastic, elasticity, and flexibility (Made Safe 2016).

14.2.4 Low-Density Polyethylene (LDPE)

Low-density polyethylene is a softer, more transparent, and more malleable HDPE variant. It is frequently used as a liner in beverage cartons, corrosion-resistant work surfaces, and other items. Examples include drinking cups, bubble wrap, sandwich and bread bags, plastic wrap, waste bags, and cling wrap. Properties: LDPE is a transparent, waxy surface that softens at 70 °C, easily scratched, and not recyclable (Guo et al. 2020).

14.2.5 Polypropylene (PP)

One of the strongest kinds of plastics available is PP. This is because it can withstand greater heat than other plastics. It is also perfect for food storage and packaging. Although it is slightly bendable due to its flexibility, it holds its strength and form for a long time. Straws, bottle caps, prescription bottles, hot meal containers, packing tape, disposable diapers, and DVD/CD boxes are some examples of PP. Its characteristics include hardness and translucence, resistance to solvents, and versatility. PP is generally a less toxic plastic, according to FDA, although its adverse effect on organisms has also been indicated (Jemec Kokalj et al. 2022).

14.2.6 Expanded Polystyrene and Polystyrene (PS or Styrofoam)

Polystyrene, often known as Styrofoam, is inexpensive and excellent at insulating, making it a standard in the building, packaging, and food sectors. Polystyrene, like PVC, is hazardous because it is easy to leach toxic substances like styrene, a neurotoxin, into the food people consume. Some of the properties are semi-toughness, glassy, rigidness, and transparency. It also softens at 95 °C; is influenced by fat, acids, and solvents; is resistant to salt solutions and alkalis; has low water absorption capacity; is clear; and is odourless and tasteless when unpigmented (Hahladakis and Iacovidou 2018). Examples include cups, takeaway containers, packaging materials for shipping and delivery, egg crates, cutleries, and insulation materials for building.

14.2.7 Polycarbonate and Other Materials

Polycarbonates and other plastic types are included in this category, especially if they fall outside one of the other six categories mentioned above. Other plastics formed from the combination of different types are also classified into this group. This plastic group cannot be recycled. Some examples include translucent plastic flatware, infant and sports bottles, gadgets, CDs and DVDs, lighting fixtures, cooler bottles, automotive and appliance parts, all resins, and multi-materials (such as laminates) with qualities reliant on plastic or a mixture of plastics. Leachates from this plastic group may result in endocrine issues in foetuses and children, such as obesity and cancer (Khan et al. 2019).

14.3 Plastics Are Classified According to Size

Plastic debris of size less than 5 mm is known as microplastic, while plastic debris of 5 mm to 20 mm and 20 mm and above is called meso debris and macro plastic, respectively. Larger plastic materials like plastic bags and fishing nets are the prime sources of macro-plastics. A substantial proportion of these plastic materials persist in the environment, contributing seriously to environmental pollution. Plastics are degraded due to exposure to ultraviolet (UV) radiation, which catalyses plastic's photooxidation, causing it to become brittle. Plastics are degraded into smaller fragments of the micro- (0.1–1000 μm) and potentially nano-sized (≤0.1 μm) particles, referred to herein as micro- and nanoplastics, respectively. When abandoned in the ocean, marine animals continuously get trapped in them (Andrady et al. 2022). Microplastics (MPs) and nanoplastics (NPs) often start as macro or meso debris, larger plastic polymers that break down into smaller pieces over time due to chemical degradation. Therefore, MPs and NPs that persist in the environment likely originate from the pre-weathering of plastics released into the biosphere.

14.4 Sources of Plastic Leaching

14.4.1 Plastic Leachate Is Released through Personal Care Products

Exposure to plastic leachate is possible through a variety of domestic and industrial applications, such as the use of plastic materials as abrasive exfoliants used in the preparation of personal care products [microbeads, alumina, sodium tetraborate decahydrate particles, and polyethylene (PE) beads] (Sun et al. 2020) and other products like hand sanitiser, soap, shower gel, cleansing products, makeup cosmetics, facial cleansers, toothpaste, and shaving cream (Nizzetto et al. 2016). Most producers of personal care products defy the use of natural materials, probably due to their high cost and availability. These natural products include pumice, oats, apricots, powdered fruit cores, or walnut shells (Decker and Graber 2012; Napper et al. 2015).

14.4.2 P*lastic* Leachates *Are R*eleased *From W*astewater *T*reatment *P*lants

Municipal wastewater contains plastic fibres from machine-washed garments that are copiously released into the environment (Wright and Kelly 2017). Most home wastewater, especially in developing nations, contains microplastics. Some microplastics can escape wastewater treatment plants (WWTPs) and seep through the land surfaces into aquatic environments or agricultural fields (Onyedikachi et al. 2018). Recently, it was noted that despite the wastewater treatment plant reduction of the microplastic content by about 98%, close to 44 million plastic leachates (microplastics) were still discharged into the receiving water per day. Also, approximately eight trillion microbeads are discharged into aquatic and other biological ecosystems every day in the United States via WWTPs, constituting a significant source of plastic pollution (Wright and Kelly 2017). Due to the discharge of industrial effluents, sludge by-products from wastewater treatment plants, and wastewater treatment plant effluents, plastic leachate has entered agricultural land, which could wind up in aquatic habitats and agricultural areas. It has been found that synthetic (plastic) clothing fibres are found in these habitats, and they persist for up to five years post-application (Ragoobur et al. 2021).

14.4.3 *Marine Debris Are Carriers of Plastic Leachate*

Plastic predominates among marine trash, including glass, metals, paper, textiles, wood, and rubber, but MPs are frequently the most prevalent (Novikov et al. 2021). MPs are pervasive and have been found in several environments around the planet, from the poles to the equator. Global sea surface pollution is estimated to be 5.25 trillion plastic particles, while the deep Indian Ocean floor is contaminated by four billion fibres per square kilometre (Sharma and Kaushik 2021). According to the fugacity study by Jang et al. (2022), marine invertebrates living on the island contaminated by marine debris accumulated more plastic concentrations than those living on the island less affected by marine debris. Therefore, marine trash contaminated with plastics are carriers and sources of leachates which are seriously injurious to health (Jang et al. 2022).

14.4.4 Plastic Leachate Released in Common Household Products

Nanoplastics are also delivered through various items, including electronics, paints, adhesives, and medical delivery systems. For example, 3D printing can release polymeric nanoparticles in a printing press. Their diminished size may bring on their potential toxicity due to environmental deterioration. The quality of the ecosystems' microbiota and the entirety of the food chain may be altered by these micro- and nano-pollutants, which must be regarded as environmental contaminants of rising concern (Garcia-Muñoz et al. 2023).

14.4.5 Plastic Leachates' Unique Properties Make Them Vectors for Various Priority Contaminants

Microplastics are implicated in negative health impacts due to their ability to traject priority pollutants listed in the Stockholm Convention as chemicals of concern (Rahman et al. 2021). Hydrophobic organic contaminants (HOCs) such as polycyclic aromatic hydrocarbons (PAHs), organochlorine pesticides, and polychlorinated biphenyls (PCBs) can be concentrated and adsorbed by microplastics due to their hydrophobic surface (Jin et al. 2020; Jiménez-Skrzypek et al. 2021). These HOCs act as chemical additives since they are adsorbed and incorporated into plastic goods. Thus, they become vulnerable to leaking into the external media because they are not chemically linked to the plastic polymer matrix (Wright and Kelly 2017). Heavy metals, including lead, cadmium, zinc, and nickel, get accumulated on the microplastics and continue breaking apart, creating a chance for inherent compounds to migrate up the surface in a concentration gradient continually. Such contaminants may be ingested and discharged into the environment, such as on agricultural lands (Onyedikachi et al. 2019a, b). If microplastics accumulate over time, they may be a source of chemical leachates in tissues, fluids, and the entire environment, resulting in serious health risks.

14.4.6 Plastic Leachate through some Agricultural Practices

Plastics are increasingly used in agricultural practices worldwide (PlasticsEurope 2018). For example, plastic mulching is often used in agriculture to maintain heat, retain water and fertiliser, enhance soil quality, and inhibit weed growth (Liu et al. 2017; Gao et al. 2021). Other applications for plastic in agriculture include plastic crates for harvesting crops, plastic crates for greenhouses, plastic fittings and spray cones, plastic irrigation pipes, plastic water storage tanks, plastic films to store silage, woven polypropylene bags of synthetic fertilisers, and compost (PlasticsEurope 2018). Unfortunately, plastics are left in agricultural areas since removing the thinner plastic films takes a lot of time and effort after the crop cycle (Steinmetz et al. 2016), hence promoting the breakdown of plastic additives which seep through the soil-water-air affecting the health of various organisms inhabiting them.

14.5 Pathways of Exposure to Plastic Pollution

14.5.1 Inhalation

Recently, reports of atmospheric dispersion of plastic material have surfaced, suggesting a potential exposure pathway through breathing (Sridharan et al. 2021). Humans are exposed to microplastics and their additives through product application and inhalation (Fig. 14.2). When building finishes, such as paints and other materials, are used, large amounts of plastic leachate are discharged (Testai et al. 2016). When plastics and plastic products are exposed to high temperatures and indiscriminately disposed of on land or burnt in the open air, toxic chemicals can leach into food, drinks, and water, releasing toxic

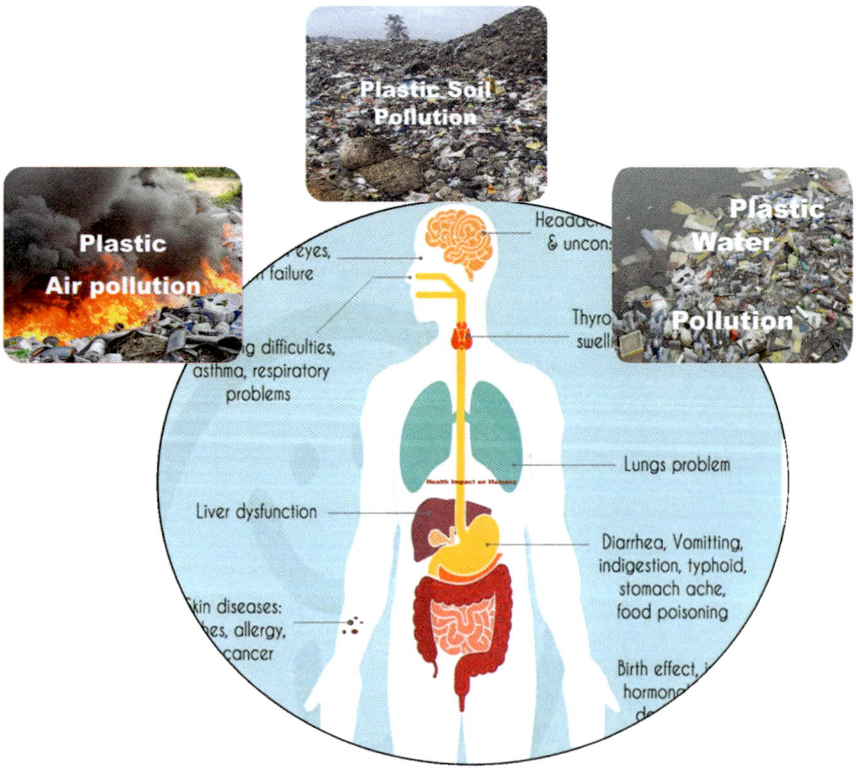

Fig. 14.2 Plastic pollution is a pathway for air, water, soil, and health pollution impact (Bansal and Sharma 2021)

chemicals into the air, inhaled into the biological system (Fig. 14.2), and resulting in public health hazards (Alabi et al. 2019). The risk to human health should be evaluated, especially the potential risk to the respiratory tract. Due to its small weight, air-driven plastic leachate can travel long distances accumulating large amounts of plastic waste before entering sewage systems. In addition, other sources of plastic pollution through the inhalation pathway include agricultural polyethylene sheets, the release of fibres from drying clothing outside, and they can also be blown from sludge-based fertiliser, which might potentially cause airborne plastic release (Koutnik 2022). Through inhalation, these plastic leachates enter biological systems and are a source of severe health problems.

14.5.2 Oral

Plastic leachates gain entry into the biological systems through the mouth, that is, through food and water consumption (Fig. 14.2). Human exposure to plastic additives like BPA is commonly anticipated to come primarily from consuming plastic-contaminated food and drinks (Kumar 2018). According to research by Vandenberg et al. (2007), groundwater can get contaminated when plastic additives like BPA are liberated from plastic polymers in landfills.

14.5.3 Dermal

The skin is a pathway through which plastic leachates can reach biological systems. They are extensively discharged using products such as electronics, building materials, toys, CDs, paints, and medical devices, which results in exposure through dermal absorption (Geens et al. 2012; Testai et al. 2016). Workers at thermal paper companies, particularly those engaged in the production of coating material and operating coating machines, as well as those exposed to thermal paper containing BPA, such as cashiers, have been shown to have dermal contact exposure to plastic leachates (Ndaw et al. 2016; Heinälä et al. 2017).

14.6 Additives Used in Plastic Production Are a Major Factor in Chemical Leaching

The most popular materials used by producers to enhance the quality of plastics are Bisphenol A (BPA), plasticisers, flame retardants, antioxidants, acid scavengers, lubricants, light and heat stabilisers, pigments and dyes, antistatic and anti inflammatory agents, slide compounds, and thermal stabilisers (Hahladakis et al. 2018; Galloway et al., 2018). Emerging priority contaminants such as BPA, polybrominated diphenyl ether (PBDE), phthalates, flame retardants, and others are not chemically bound to the plastic polymer and hence possess the tendency to migrate within the substance, reach its surface, and then leach out into the environment. Leachates are mixes of additives, some of which are developing pollutants or those that might be dangerous for the environment and people's health (Gunaalan et al. 2020). Plastic additives are usually added to plastics to improve their softness, harden them, or add colour, which has generated concerns. Plastics break down into little pieces, and addivtives leak into meals, bodies of water, and the environment (Gunaalan et al. 2020). Numerous of them have been connected to harmful health impacts, ranging from disturbing our sensitive endocrine system to problems in reproduction. Obesity, early puberty, and adult onset diabetes have all been connected to exposure to chemical leachates. These additives are invariably the most expensive components of plastic formulations and the minimum quantity required to produce a plastic product. However, the endocrine system may be modulated if the biological system is exposed to these plastic leachates. They are thus classified as EDCs, or endocrine-disrupting compounds (Andersen 2019).

14.6.1 Effect of Bisphenol a

Bisphenol A is a synthetic chemical that is widely used in the production of tough, long-lasting plastics like polycarbonate and epoxy resins. In the manufacturing of polyvinyl chloride (PVC), it also functions as a stabiliser. BPA in the plastic product might disintegrate and leak into any food it comes into contact with. Exposure to BPA is critical and common. Research indicated that it is found in the urine in more than 90% of Americans (Ye et al. 2015). Also, research has demonstrated that BPA is a systemic toxicant even at low concentrations and polymerises to produce a hard plastic (Kumar 2018). The health risk implication brought on by BPA exposure has been challenging to manage. Hence, the "European Human Biomonitoring Initiative" (HBM4EU) recently developed human biomonitoring guiding values (HBM-GVs). An HBM-GV is a biomarker concentration in the biological matrix and serves as a guideline value below which unfavourable human health consequences caused by drug exposure are not anticipated (Apel et al. 2020). Increased temperature, duration of contact, and pH may trigger hydrolysis and breakdown at the polymer surface, which might lead to BPA being released from polycarbonate (Pedersen et al. 2015). Despite its widespread use and manufacture, few studies have looked into BPA occupational hazards in Europe (Ribeiro et al. 2017). These substances are often leached into water or food through container lining in pregnant women's blood, amniotic fluid, placental tissue, and cord blood, indicating foetal exposure to BPA. According to Sayıcı et al. (2019), BPA can also be passed from mothers to their infants through breast milk or feeding bottles. Due to the growing concerns of BPA accumulation in organs and tissues and also its reduced ability to be detoxified, pregnant women are more at risk than other adults who are subjected to the same levels of exposure (Sayıcı et al. 2019). Generally, the main route of BPA exposure is through food consumption (Kumar 2018). Although, there was a belief that bisphenol-A at low amount is safe since it can break down however, a recent study has projected its potent toxicity in the body as this chemical additives can be biotransformed into a compound that may induce metabolic diseases like obesity (Kumar 2018).

14.6.2 Phthalates

Many consumer items use phthalates as plasticisers, but because they are not covalently attached to plastic, they can migrate or seep out, exposing people to them (Kumar 2018). They are quickly digested, have a short half-life (hours), and are eliminated in urine and faeces. Flexible vinyl, which is utilised in consumer items, flooring and wall coverings, food contact applications, and medical equipment, is created from plastics by adding higher molecular weight phthalates (Wright and Kelly 2017). Lower molecular weight phthalates are used in personal care goods like coatings, solvents, lacquers, and varnishing, as well as in some medications to offer scheduled releases (Kahn et al. 2020). For the general population, exposure by ingesting, inhalation, and skin contact are all regarded as significant exposure pathways. Phthalates metabolise quickly, do not build up, and are primarily eliminated through urine. According to Bamai et al. (2016), children had greater daily phthalate intakes than their parents based on estimates from urine metabolites. Due to their hand-to-mouth activity, babies and toddlers consume up to ten times more household dust than adults, making floor dust one of the most significant sources of phthalate consumption in these age groups. Studies have revealed that the levels of maternal pre- and post-natal phthalate metabolites varied greatly and were often not substantially correlated. These connections have been incredibly understudied in humans to date compared to the enormous body of evidence in laboratory animals confirming phthalate reproductive or developmental harm (Rahman et al. 2021). Phthalates have been labelled as reproductive toxins in humans due to research demonstrating altered male genital development and decreased semen quality (Kumar 2018). Males have only been included in the majority of human research examining reproductive or developmental health consequences linked to prenatal or newborn exposure to phthalates. They have concentrated on pronounced modifications such as hypospadias; shortened anogenital distance (a sensitive and non-invasive indicator of potential androgen deficiency during foetal development); and malformations of the epididymis, vas deferens, seminal vesicles, and prostate (the "phthalate syndrome"). To produce congenital defects, foetal testosterone production must be reduced during the key period for the development of these structures (Kumar 2018). Phthalates primarily change the gene expression of several enzymes and transport proteins involved in typical testosterone production and transport in the foetal Leydig cell, which impairs the development of androgen-dependent structures. This phenomenon closely matches the condition of human testicular dysgenesis (Kumar 2018).

14.6.3 Brominated Flame Retardants

Brominated flame retardants (BFRs) are synthetic less flammable chemicals made from various materials for human usage. They are used by industries for the production of plastic textiles and electrical/electronic equipment. BFRs are distributed into five main classes:

(1) Polybrominated diphenyl ethers (PBDEs) for plastics, textiles, circuitry, and electronic castings; (2) hexabromocyclododecanes (HBCDDs) used by building industries for thermal insulation; (3) tetrabromobisphenol A (TBBPA) and other phenols used as thermoplastics in TVs and printed circuit boards; (4) polybrominated biphenyls (PBBs) for plastic foams, consumer appliances, and textiles; and (5) other brominated flame retardants (OBFR) used for various insulation purposes. The European Union has restricted the use of certain BFRs due to their persistence in the environment and the risks these chemicals pose to public health. This is because BFR-treated products leach BFRs into the environment and contaminate the air, soil, and water. These contaminants may then enter the food chain, disrupting the health and well-being of man and his environment (EFSA) 2021). All plastic materials, from waiting room chairs to intravenous pumps, must be fire-resistant. Manufacturers of healthcare items must include flame-resistant chemicals, or "flame retardants", in their products in order to comply with fire safety regulations. BFRs, a subgroup of these flame retardants, are currently under close examination because of mounting evidence that they bioaccumulate in the food chain and human bodies and have negative effects on children through processes that target various levels of the hypothalamic-pituitary-gonad/thyroid axis, Thus, exposing them to certain plastic additives may change endocrine function. As a result of the generated molecular epigenetic changes, many organs may experience transgenerational effects (Kumar 2018).

14.7 Soil Physical and Chemical Properties Influencing Plastic Leaching

The characteristics of the soil determine how badly microplastics harm terrestrial ecosystems and human health (Lehmann et al. 2019; Liu et al. 2017). Changes in soil bulk density, water-holding capacity, soil aggregate stability, and soil water repellency occur when soils are exposed to microplastics. Microplastics can change soil bulk density, an important factor in predicting soil carbon storage, because they generally have a lower density than soil particles. In soils, microplastics in the form of plastic film have the ability to change the porosity, which would hasten water evaporation and cause soil cracking. Studies have shown that the use of polyester fibre can greatly improve the soil's capacity to store water while drastically reducing the proportion of water-stable aggregates in the soil.

Soil enzymes assist in the breakdown of organic matter and the cycling of different elements (C, N, P, etc.). (Trasar-Cepeda et al. 2008; Allison and Jastrow, 2006). Urease and phosphatase are engaged in the cycle of nutrients in the soil and act as indicators of soil quality during times of stress (Hagmann et al. 2015; Xiao et al. 2017). Fluorescein diacetate hydrolase (FDAse), a significant indicator of soil microbial metabolic activity, may quickly reflect changes in soil quality (Muscolo et al. 2015). Fei et al. found that 1% and 5% (w/w) additions of PVC and PE to soils reduced FDAse activity and increased urease and acid phosphatase activity. Although PE microplastics had a greater impact on urease activity in the soil, PVC microplastics had a greater impact on acid phosphatase and FDAse activities in the soil (Fei et al. 2020). More proof was presented by Yi et al. (2021) showing the kind and shape of microplastics that affected the activity of soil enzymes. Recent studies, however, discovered that introducing soil microplastics at lower concentrations [0.2 and 1.0% (w/w)], which are more indicative of their real amounts in the soil environment, had no appreciable influence on the activities of soil enzymes (Xu et al. 2020).

In addition, microplastics have an effect on the soil nitrogen cycle. Recent studies have shown that the carbon and nutrient cycles may be affected by soil-dissolved organic matter (DOM), which is altered by microplastics (Liu et al. 2017). Ecosystems on farms have received increased attention recently since they account for 20% of global greenhouse gas emissions (Verge et al. 2007). Microplastics in the soil might change microorganism-controlled soil greenhouse gas emissions (Liu et al. 2019; Liu et al. 2017). According to research by Gao et al., soils with 18% microplastics might increase soil CO_2 emission flow by 28.67%. CO_2 emissions were also found to have a strong positive correlation with microplastic-resistant microbial species such as Mycobacterium, Aeromicrobium, and Amycolatopsis (Gao et al. 2021).

Nitrogen is a nutritional element that is necessary to support a number of significant biological processes in soils, including nitrification, ammonification, and denitrification. Excess plastic mulch may reduce soil inorganic nitrogen, according to long-term plastic mulching studies. The influence of polyethylene microplastics on the nitrogen cycle was established by the rise in urease activity in their presence (Huang et al. 2019). Polystyrene and polyethylene microplastics also inhibited leucine aminopeptidase and N-acetylglucosaminidase, two critical enzymes impacting the soil nitrogen cycle (Awet et al., 2018; Bandopadhyay et al. 2019).

14.8 Health Issues Associated with Plastic Leaching

14.8.1 Humans

The health concerns associated with plastic polymers are often thought to be caused by various sorts of plastic additives and leftover monomers that are apparently retained from these polymers, despite the fact that they are widely believed to be safe and barely endanger society. According to Fucic et al. (2018), the majority of plastic additives are recognised endocrine disruptors and carcinogens, and they damage people mostly when they come into contact with their skin (leading to dermatitis), mouth (swallowing), or by inhalation (Aalto-Korte et al. 2019). When ingested by a variety of marine and freshwater species, microplastics are important poisons that can create complexes in the food chain and cause severe health issues (Wright and Kelly 2017). Animals exposed to plastic additives and microplastics can be detrimental to people if consumed in food. Biomonitoring studies on human tissues have shown that plastic components are present in the human species through the detection of environmental pollutants (Smith et al. 2018). The biological system exposure to plastic leachates is associated with diabetes, heart disease, and several health concerns (Gopinath et al. 2022). In foetuses, newborns, children, and pregnant women, high levels of this endocrine disruptor (BPA) have a deleterious impact on the brain, behaviour, and prostate gland (which may result in foetal or neonatal death, congenital impairments, or lower birth weight in their kids) (Kahn et al. 2020). Additionally, studies revealed that both men and women's free and total testosterone levels had high levels of BPA (Kumar 2018). The idea that the human body can break down BPA into a harmless product was refuted in a study by (Kumar 2018), since the bi-product BPA-glucuronide (BPAG) stimulates adipogenesis and may significantly contribute to obesity.

14.8.2 Aquatic Organisms

Diverse aquatic organisms, including coral, polychaete worms, sea cucumbers, crustaceans, molluscs, fish, reptiles, water birds, and sea mammals, all contain plastic leachates. Some of these species are able to excrete or eat plastic waste, while others retain, accumulate, and immobilise it in their bloodstream (Anderson et al. 2017; Li et al. 2019; Maaghloud et al. 2020). The three mechanisms that make up the plastics' toxicity seem to be (1) ingesting stress (physical obstruction, energy expenditure during egestion); (2) additive leaks from plastic (plasticisers); and (3) exposure to contaminants linked to microplastics (such as persistent organic pollutants) (Anderson et al. 2017). By changing light penetration into the water column and sedimentation properties, plastics can also have an impact on the environment's abiotic attributes (Eerkes-Medrano et al. 2015). Recent studies on plastic leachates have mostly focused on aquatic creatures. Both micro- and macroplastics have an influence on larger creatures (such as large fish, reptiles, birds, and mammals), but microplastics primarily have an impact on smaller organisms (such as zooplankton, worms, coral, crabs, molluscs, and tiny fish). However, because they are commensals whose primary victims are typically the smaller creatures in aquatic environments, bigger species in the water environment are more impacted by microplastics than macroplastics. Macroplastics have a detrimental impact on these species by restricting actions in vertebrates, including mammals, reptiles, and aquatic birds (such as swimming, breathing, and eating), lowering their ability to survive, and impeding growth and reproduction (Beer et al. 2018). According to Ramesh et al. (2019), the entanglement of fish nets and plastic waste discharged into the water had a significant negative impact on two sea turtles. The fish's arm tissues were ripped, and a person died as a result. Some turtles eat macroplastics because they mistake plastics for food and become caught in fishnets and huge pieces of plastic. This makes it difficult for them to eat and conceal from predators (Li et al. 2019; Nelms et al. 2016). The presence of plastic particles in the digestive systems of several seabirds, fish, and mammal species from tropical, temperate, and polar climates is linked to the polymer's attachment to the exterior surface, which obstructs and harms the digestive system. Inflammation, hepatic stress, and reduced development are further impacts (Auta et al. 2017). The clogged gut and cloaca, respectively, also have an impact on the female's capacity to reproduce (Nelms et al. 2016; Li et al. 2019). The leaching of contaminants, such as trace metals and other toxins (such as persistent organic pollutants), through the plastics into the digestive tracts would result in developmental and reproductive abnormalities in animals. This is one of the secondary effects of macroplastic consumption in large animals (Nelms et al. 2016). Plastics on beaches also cause the sand temperature to drop, which has a significant impact on the change in sex ratios of reptiles (such as turtles) that lay their eggs on beaches (Nelms et al. 2016). There are several ways that microplastics might infiltrate the aquatic biota, including filter feeding, suspension feeding, eating of prey that has been exposed to microplastics, or direct ingestion (Anderson et al. 2017). Microplastics can bind to organic pollutants such as polycyclic aromatic hydrocarbons, polybrominated diphenyl ethers, polychlorinated biphenyls, and dichloro diphenyl trichloroethane because they are hydrophobic and have vast surface areas (Anderson et al. 2017). In aquatic environments, plastics could act as a substrate for bacterial colonies and as vectors for infections (Anderson et al. 2017). Microplastics therefore affect organisms on a wide range of levels, including genetic structure and expression; biochemical activities (such as immune response, oxidative and energy-related enzyme activities); behavioural changes (such as swimming, feeding, olfactory senses, inflammatory responses, and other daily activities); alteration of life history traits (such as development, survival, reproduction, size, and weight); and health impairment such as lung problems, liver dysfunction and skin diseases (Auta et al. 2017; Barboza et al. 2018).

14.8.3 Soil Organisms

In recent years, plastic leachates have become pervasive pollutants that are a source of growing worry due to their toxicity and other health impacts, not just in aquatic and terrestrial ecosystems. Despite the fact that soil is a significant reservoir and a means of microplastic leaching and spreading to other ecosystems (Monkul and Özhan 2021), microplastics may alter the ecosystem where soil creatures live, indirectly impacting such species (Selonen et al. 2020); and this connotes a rippling effect that in turn affects other organisms. The effects of microplastics on earthworms have received the most significant attention since they are a model species in the soil ecosystem and their development rates and immune systems are impacted by the types and quantities of microplastics (Hodson et al. 2017). According to research by Cao et al. (2017), earthworm adaptation had no discernible effects when exposed to 0.25–0.5% of polystyrene microplastics. Growth inhibition only occurred at exposure concentrations higher than 1% Cao et al. (2017). Earthworm mortality rose by 8% and 25%, respectively, when exposed to microplastic concentrations of up to 28% and 60% (Lwanga et al. 2016). Microplastics have also been shown to alter the intestinal microbiota of soil organisms, suggesting that these communities may be involved in the cycling of critical elements and the consumption of organic materials (Zhu et al. 2018a). Landfills, soil supplements, agricultural films, tyre abrasion, and atmospheric deposition are the main entry points for microplastics into the soil ecosystem. Microplastics incorporated into the soil may change its composition and interact with other soil components, impacting how well the soil functions and how many organisms exist (Wang et al. 2019). Most of the time, microplastics are accidentally consumed by species that cannot break down the plastic into smaller pieces (Wang et al. 2019). According to studies conducted on marine organisms (Setälä et al. 2016; Lahive et al. 2019), the ingestion of MPs may cause growth loss, reproduction reduction, and mortality of terrestrial organisms due to nutritional imbalance, organ damage, and disorders of immune responses and metabolisms (Hodson et al. 2017; Wang et al. 2019). Furthermore, as humans are the top predators in the terrestrial food chain, the bioaccumulation and subsequent trophic transfer of MP leachates might potentially harm their health (Guo et al. 2020) The composition and diversity of the microbial population, notably of rhizosphere microorganisms like N-fixers, mycorrhizal fungi, and pathogens, might be impacted by the presence of MPs, which modify soil structure and function (Rillig et al. 2019). Numerous studies have shown that the presence of MPs has an impact on a variety of soil microbial enzymes, including dehydrogenase, leucine-aminopeptidase, alkaline phosphatase, glucose-glucosidase, cellobiohydrolase, and fluorescein diacetate hydrolase. The microbiota that resides in their stomachs is also impacted by microplastics, in addition to the organisms that dwell in the soil. *Caenorhabditis elegans*' body length, survival rate, reproduction rate, and genes for oxidative stress were all shown to be influenced by the size of polystyrene (PS). PS might also significantly alter these characteristics (Lei et al. 2018). In addition to affecting the bacterial diversity and microbiota in the gut of collembolan (*Folsomia candida*), Zhou et al. (2020) found that exposure to 0.1% PVC microplastics reduced the growth and reproduction of collembolan (Zhu et al. 2018b). The consumption of PETs by snails induced changes in their TAOC (comprehensive index representing oxidative stress), GPx antioxidant enzyme activity, and malondialdehyde (MDA) concentration. These changes resulted in lipid peroxidation and digestive system impairment in the snails (Song et al. 2019). Snails ingesting infected leaves may modify the histology of their digestive systems as well as the growth, behaviour, and viability of the microbiota that inhabits their intestinal tracts (Chae and An 2020).

14.9 Plastic Pollution and Sustainable Development Goals: A Nexus for Human and Economic Development

Plastic leachates into foods, water bodies, and agricultural lands through various routes can impair human and economic development in connection to the SDGs (Fig. 14.3), and it is a growing concern globally:

1. The well-being and health of man and other organisms. The poisonous particles can cause local inflammation and all kinds of physiological effects which should be combated (Fig. 14.3) to prevent the health concerns associated with plastic pollution (SDG 3).
2. Clean water and sanitation are essential for the existence of man and other living things; for example, without clean water free from contaminants, living organisms are exposed to different disorders (Fig. 14.3). Contaminated water affects the growth of plants and also affects the organisms in the soil whose activities enhance plant growth and degrade pollutants (SDG 6). This is aimed at promoting the quality of freshwater by remediation processes and reduction in the plastic pollution loads.
3. Sustainable cities and communities (SDG 11): By the effective collection and processing of waste, reusability, and ensuring clean-up of our cities and communities especially in densely populated regions where plastic waste is abundant, hence avoiding health risks associated with plastic pollution.
4. Responsible consumption and production (SDG 12): The abundant production of plastics for different purposes and the single-use practice, that is, non-recycling of plastic

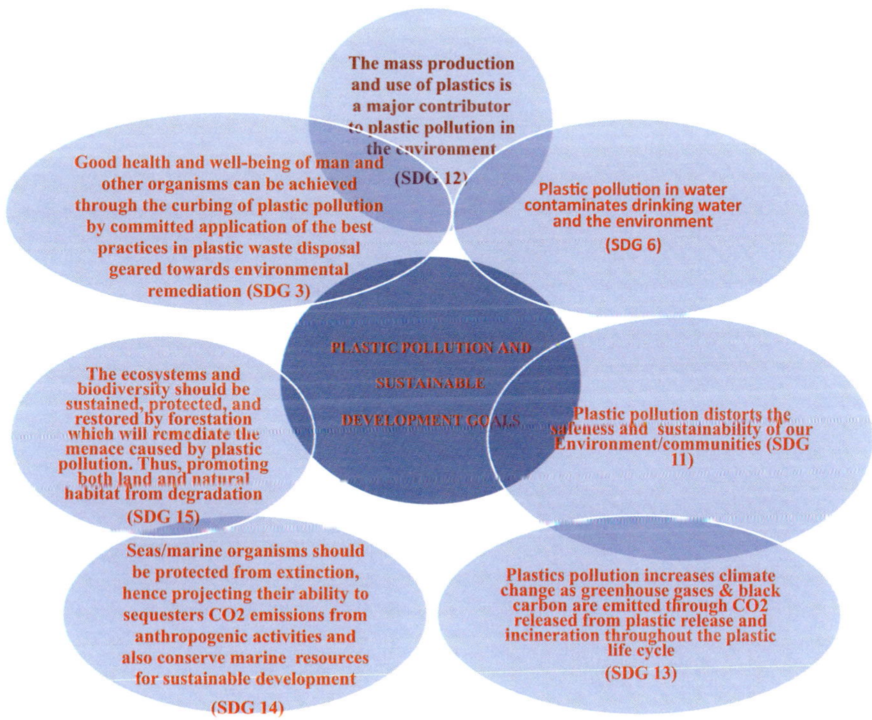

Fig. 14.3 The relationship between plastic pollution and target sustainable development goals (SDGs) (United Nations 2015)

materials, are major contributors to plastic pollution in the sea and on land. This pollution has a negative influence on the functioning of ecosystems and endangers animal lives as well as the food supply of large groups of people. Also, the consumption of plastics via several exposure pathways has promoted the observable link in health challenges (Fig. 14.3) caused by plastic pollution. Hence, the best way to promote SDG 12 is through an absolute reduction in plastics, improved effectiveness of recycling, and safer alternatives.

5. Climate Action (SDG 13): Most plastic materials are made from fossil fuels which use lots of energy and when incinerated cause global warming and the release of carbon dioxide (Fig. 14.3). The reduction of CO2 emissions would prevent an average temperature increase of two degrees is an extremely urgent SDG goal.
6. Protection of seas and oceans (SDG 14): This SDG is aimed at reducing and preventing pollution in the sea, especially from waste that originates on land (Fig. 14.3). This is because 80% of marine waste comes from the land and most of the waste dumps and drainages are infested with plastics which find their way in the water bodies. Huge benefits can be reaped if effective waste collection systems are put in place everywhere to assuage plastic pollution.
7. Repair ecosystems and retain biodiversity (SDG 15): Ecosystems in the sea and on land are threatened by plastic leachates which seep through into the ecosystems, causing severe diversity losses and suffocation of animals, making them unable to feed well, and also endangering their species and health.

14.10 Remedy for Plastic Leaching Health Effect

14.10.1 Bioplastics

Due to the mounting environmental strain on global warming and plastic waste, bioplastics have drawn a lot of attention. One of them, poly (lactic acid) (PLA), is both bio-based and biodegradable and has been extensively employed in numerous applications involving throwaway/takeaway packaging. According to Jem's law, the demand for PLA in the worldwide market doubles yearly when compared to the demand for conventional petroleum-based plastics. PLA often has fewer mechanical and physical qualities and is more costly. PLA's mechanical and thermal properties might be enhanced for high-end applications by recent compounding work and commercialisation of D-lactic acid and its polymer, PDLA (e.g. by making stereocomplex PLA). PLA is still used in other applications, although. Poly (glycolic acid) (PGA), which has a structure comparable to PLA and promising qualities including strong biodegradability and barrier properties, may be a useful addition to PLA. Co-polymerisation, physical mixing, and multilayer lamination can all be used to modify PLA with PGA. PGA and its combination with PLA have received a lot of attention in biomedical applications, but because of their relatively expensive manufacturing costs, they have not been extensively developed at a large scale. The emergence of new government rules and the development of novel manufacturing technologies are, in this case, the primary forces behind a global shift towards bioplastics. Numerous government rules that limit the use of conventional plastics and promote the use of biodegradable polymers have recently been published. Industrial waste gases may be converted into PGA using cutting-edge production equipment, which lowers production costs and carbon emissions. PGA and PLA may be coupled to play a crucial role in the sustainable and ecologically friendly plastic industry by advancing the production and compounding technologies, especially for single-use items needing quick disintegration at room temperature or in the natural environment.

14.10.2 Alternatives to Plastics

Although it is hard to imagine living without plastics, the hunt for alternatives has been sparked by growing knowledge of the health risks. On the horizon, several excellent options are:

1. Increasing the use of glass materials, notably for storage and microwave use.
2. Prefer paper, jute, or cloth-based reusable bags over plastic ones.
3. Compared to their petrochemical equivalents, the use of bioplastics, which are largely produced from sustainable resources like biomass, can result in production energy savings of up to 40%.
4. Biodegradable plastics include those made of polyhydroxyalkanoate (PHA) polyesters, starch-based polymers, and aliphatic polyesters.
5. A promising material is one manufactured from milk casein and chicken feathers.
6. Produced from biodegradable lignin, a by-product of the paper industry, is liquid wood.
7. By ensuring the protection and enhancement of the environment, avoiding environmental pollution in all its forms, and addressing special environmental issues unique to different nations, government initiatives programmes might support the strict reforms in plastic waste management.

14.10.3 Plastic Recycling

Recycling of plastic materials like bags and containers instead of single-use practices often used in packaging can reduce the adverse effects of plastic pollution and packaging costs, as waste materials can find a secondary route to obtain value rather than being disposed of. This will reduce dumping in oceans and rivers, and as well protect marine organisms and asides reduce global warming which are important goals in the world leaders' SDG plans. Restaurants, stores, and events can purchase, use, refill, and wash them hundreds of times, before returning them to recycling or reshaping when they are not needed anymore. This is one of the preferred routes as less energy and resources are required to carry it out. However, reused plastic packaging may be contaminated from their former use. hence, to avoid migration of contaminants to food, it is important to wash first before reusing plastic packaging to ensure health safety (Ncube et al. 2021). Examples of plastic recycling products and methods include Waterhaul, used in making sunglasses from plastic waste; WasteAid, a Gambian initiative for converting LDPE plastic waste to plastic tiles; Kkplascticroads, an Indian initiative using recycled plastic waste as binders in making asphalt roads; and Plastic Fantastic, an initiative that uses litters to reprocess plastic litters into new products for scaffolds, benches, and facing materials. Plastic collective shrudder is also equipment for recycling plastic waste, and the Byfusion method also reuses plastic in making building materials (Plastic soup foundation 2023). These beautiful recycling processes actually go a long way to reduce the accumulation of plastic in the environment. This chapter encourages its application in achieving SDG goals come 2030.

14.10.4 Technological Tools for Detecting and Cleaning up Plastics

Technology tools for detecting and cleaning up include the following: (a) the detection method involves: (1) WASSER 3.0 PE-X, a project developed for the removal of microplastics and micropollutants from water using hybrid silica gel that binds to plastic materials thus forming lumps that can be separated easily on the surface of the treatment basin; (2) MAGNET, developed by some Spanish scientists which uses a system of hybrid material that absorbs microplastics from water. It can be used to clean small samples and monitoring of water pollution; (3) DRAPER, developed by an American company which uses fluorescent dye on water samples from the sea, making plastics easily identifiable around small organisms such as plankton and organic material; and (4) Marine Litter Detective, a floating tracker giving insight into the sea currents that carry plastic along (Plastic Soup Foundation 2023).

(b) The cleaning up process involves: (1) SeeHamsters, which are compact catamarans with a length of about 4.5 meters and a width of 2 m and a low draught, equipped with fold-down nets or fishing gear to collect debris from inland waters such as lakes and rivers. (1) the Versi-Cat's function is the collection of plastic litter and debris from the water surface to a removable basket, which can be lifted and tipped directly into a skip or shoreside receptacle for disposal. (3) the Aqua Pod are waste-collecting modular floating docks of 2.4 × 6 meters that capture plastic waste from marinas and other waterfront areas by creating a water flow into a net, thereby trapping waste for easy removal. (4) Clear Rivers is also another initiative that uses Litter Traps to collect plastic waste from rivers and port areas and use it to produce platforms that can be assembled to form a floating park (Recycled Islands). (5) the Cleaning Drone V1 is a fully electric, autonomous surface vehicle designed to remove marine plastic waste in and just below water surfaces in ports, canals, rivers, river mouths, and other marine and aquatic environments. It is designed as a robust catamaran with a collection unit and a unique system for self-emptying in a specially designed stationary waste pool. (c) Microplastic removal system (1) a static charge filtration screen used to remove microplastic from beaches. The screens are loaded with dry sand and plastic, and through a back-and-forth motion users manually filter the sand through the screen. (2) Hoola One is a machine that uses a buoyancy separation method to recover plastic particles as small as 50 um. It is made out of three different modules to allow access to hard-to-reach places, and it uses a separation technology to put natural matter back on the beaches. (3) Vroom suction installations provide suction technologies that can be used to remove waste and pollutants from the soil, sand, gravel, and water. It is very useful to remove green and residual waste from hard-to-reach places. (4) The Marina Trash Skimmer collects trash and removes oil and floating debris from the surface of the water. It works with the natural currents of the installation sites, tides, and prevailing winds to collect trash and oil sheen into one easy-to-access location, for quick removal and disposal. (5) The Manta is a plastic-eating catamaran that collects trash on an industrial scale, up to 3 tons of ocean garbage per hour. It has nets along the stern to collect plastic and garbage, along with sustainable energy sources like solar panels and wind turbines to power the onboard collection and recycling centre. (6) Eco-Mobile Robot is developed to clean up the ocean on a big scale. The floating deck, looking like an oil platform, can suck up trash. Connected to satellites, they can be monitored and controlled (Plastic Soup Foundation 2023).

(c) Natural organisms like *Aspergillus tubingensis*, a mould that enhances quicker breakdown of polyester polyurethane; *Phanerochaete chrysosporium*, contributes to the breakdown of polycarbonate (PC); and *Galleria mellonella*, a caterpillar that has the ability to eat and digest this kind of plastic (Plastic Soup Foundation 2023).

14.11 Conclusion

There is a great need to use healthy alternatives that are biodegradable in order to curtail the adverse health implications as a result of exposure to plastics. The effects of plastic additives, especially their disruptive effects on the endocrine glands, intestinal obstructions, injuries, and deaths, are a severe concern. Although plastics and their additives have significant uses that assist daily life, their environmental and significant health risks outweigh their benefits. The only way out is to look for natural alternatives, a more efficient plastic recycling process, discourage the use of plastic materials in building roads and other infrastructure, and reduce usage in the food and beverage industry. Governments, stakeholders, and technocrats should make policies that control plastic usage and its indiscriminate release into the environment in industrial and anthropogenic activities.

References

Aalto-Korte K, Suomela S, Pesonen M. Allergic reactions to lower concentrations of nickel sulfate and formaldehyde often appear later than reactions to higher concentrations. Contact Dermatitis. 2019;80(3):162–5.

Ahmed IB, Nwaichi EO, Ejikeme U, Ugbebor JN, Arokoyu SB. Cost reduction strategies in the remediation of petroleum hydrocarbon contaminated soil. Open Res Afr. 2022;5(21):1–17.

Alabi OA, Ologbonjaye KI, Awosolu O, Alalade OE. Public and environmental health effects of plastic wastes disposal: a review. J Toxicol Risk Assess. 2019;5(021):1–13.

Allison SD, Jastrow JD. Activities of extracellular enzymes in physically isolated fractions of restored grassland soils. Soil Biol Biochem. 2006;38(11):3245–56.

Andersen HM. Effects of plastic additives on precision-cut liver slices (PCLS) from Atlantic cod (Gadus morhua). (Master's thesis, Norwegian University of Life Sciences, Ås); 2019.

Anderson PJ, Warrack S, Langen V, Challis JK, Hanson ML, Rennie MD. Microplastic contamination in Lake Winnipeg, Canada. Environ Pollut. 2017;225:223–31.

Andrady AL, Law KL, Donohue J, Koongolla B. Accelerated degradation of low-density polyethylene in air and in sea water. Sci Total Environ. 2022;811:151–368. https://doi.org/10.1016/j.scitotenv.2021.151368.

Apel P, Rousselle C, Lange R, Sissoko F, Kolossa-Gehring M, Ougier E. Human biomonitoring initiative (HBM4EU)—strategy to derive human biomonitoring guidance values (HBM-GVs) for health risk assessment. Int J Hyg Environ Health. 2020;230:113622.

Auta HS, Emenike CU, Fauziah SH. Distribution and importance of microplastics in the marine environment: a review of the sources, fate, effects, and potential solutions. Environ Int. 2017;102:165–76.

Awet TT, Kohl Y, Meier F, Straskraba S, Grün AL, Ruf T, Jost C, Drexel R, Tunc E, Emmerling C. Effects of polystyrene nanoparticles on the microbiota and functional diversity of enzymes in soil. Environ Sci Eur. 2018;30(1):1–10.

Bamai Y, Araki A, Kawai T. Exposure to phthalates in house dust and associated allergies in children aged 6-12 years. Environ Int. 2016;96:16–23.

Bandopadhyay S, Sintim H, DeBruyn J. Structural and functional responses of soil microbial communities to biodegradable plastic film mulching in two agroecosystems. bioRxiv. 2019; 650317.

Bansal M, Sharma JG. Plastic pollution by COVID-19 pandemic: an urge for sustainable approaches to protect the environment. J Pure Appl Microbiol. 2021;15(3):1083–94.

Barboza LGA, Vethaak AD, Lavorante BR, Lundebye AK, Guilhermino L. Marine microplastic debris: an emerging issue for food security, food safety and human health. Mar Pollut Bull. 2018;133:336–48.

Beer S, Garm A, Huwer B, Dierking J, Nielsen TG. No increase in marine microplastic concentration over the last three decades—a case study from the Baltic Sea. Sci Total Environ. 2018;621:1272–9.

Cao D, Wang X, Luo X, Liu G, Zheng H. IOP. Effects of polystyrene microplastics on the fitness of earthworms in an agricultural soil. IOP Conf Ser Earth Environ Sci. 2017;61:012148.

Chae Y, An Y. Nanoplastic ingestion induces behavioral disorders in terrestrial snails: trophic transfer effects via vascular plants. Environ Sci Nano. 2020;7:975–83.

Chang X, Fang Y, Wang Y, Wang F, Shang L, Zhong R. Microplastic pollution in soils, plants, and animals: a review of distributions, effects and potential mechanisms. Sci Total Environ. 2022;850:157857.

Cohen S. Waste management in New York City, Hong Kong, and Beijing. In: The sustainable city. Columbia University Press; 2018. p. 109–30.

Decker A, Graber EM. Over-the-counter acne treatments: a review. J Clin Aesthet Dermatol. 2012;5(5):32–40. PMID: 22808307; PMCID: PMC3366450.

Eerkes-Medrano D, Thompson RC, Aldridge DC. Microplastics in freshwater systems: a review of the emerging threats, identification of knowledge gaps and prioritisation of research needs. Water Res. 2015;75:63–82.

Eriksen MK, Christiansen JD, Daugaard AE, Astrup TF. Closing the loop for PET, PE and PP waste from households: influence of material properties and product design for plastic recycling. Waste Manag. 2019;96:75–85.

European Food Safety Authority (EFSA). Administrative guidance for the preparation of applications on recycling processes to produce recycled plastics intended to be used for manufacture of materials and articles in contact with food, vol. 18(3). 2021. p. 6512E.

Fei Y, Huang S, Zhang H, Tong Y, Wen D, Xia X, et al. Response of soil enzyme activities and bacterial communities to the accumulation of microplastics in an acid cropped soil. Sci Total Environ. 2020;707:135634.

Fucic A, Galea KS, Duca RC, El Yamani M, Frery N, Godderis L, et al. Potential health risk of endocrine disruptors in construction sector and plastics industry: a new paradigm in occupational health. Int J Environ Res Public Health. 2018;15(6):1229.

Galloway TS, Lee BP, Buric I, Steele AM. Plastics additives and human health: a case study of bisphenol A (BPA). Plast Environ. 2018;17:131.

Gao B, Yao H, Li Y, Zhu Y. Microplastic addition alters the microbial community structure and stimulates soil carbon dioxide emissions in vegetable-growing soil. Environ Toxicol Chem. 2021;40:352–65.

García-Muñoz P, Robert D, Ruppert AM, Keller N. Microplastics (MPs) and nanoplastics (NPs): introduction. In: Current developments in biotechnology and bioengineering. Elsevier; 2023. p. 1–32.

Geens T, Neels H, Covaci A. Distribution of bisphenol-A, triclosan and n-nonylphenol in human adipose tissue, liver and brain. Chemosphere. 2012;87(7):796–802.

Gopinath PM, Parvathi VD, Yoghalakshmi N, Kumar SM, Athulya PA, Mukherjee A, Chandrasekaran N. Plastic particles in medicine: a systematic review of exposure and effects to human health. Chemosphere. 2022;303:135227.

Gunaalan K, Fabbri E, Capolupo M. The hidden threat of plastic leachates: a critical review on their impacts on aquatic organisms. Water Res. 2020;184:116170. https://doi.org/10.1016/j.watres.2020.116170.

Guo JJ, Huang XP, Xiang L, Wang YZ, Li YW, Li H, Cai QY, Mo CH, Wong MH. Source, migration and toxicology of microplastics in soil. Environ Int. 2020;137:105263. https://doi.org/10.1016/j.envint.2019.105263.

Hagmann D, Goodey N, Mathieu C, Evans J, Aronson M, Gallagher F, et al. Effect of metal contamination on microbial enzymatic activity in soil. Soil Biol Biochem. 2015;91:291–7.

Hahladakis JN, Iacovidou E. Closing the loop on plastic packaging materials: what is quality and how does it affect their circularity? Sci Total Environ. 2018;630:1394–400.

Hahladakis JN, Velis CA, Weber R, Iacovidou E, Purnell P. An overview of chemical additives present in plastics: Migration, release, fate and environmental impact during their use, disposal and recycling. J Hazard Mater. 2018;344:179–99.

Heinälä M, Ylinen K, Tuomi T, Santonen T, Porras SP. Assessment of occupational exposure to bisphenol A in five different production companies in Finland. Ann Work Exposures Health. 2017;61(1):44–55.

Hodson M, Duffus-Hodson C, Clark A, Prendergast-Miller M, Thorpe K. Plastic bag derived-microplastics as a vector for metal exposure in terrestrial invertebrates. Environ Sci Technol. 2017;51(8):4714–21. https://doi.org/10.1021/acs.est.7b00635.

Huang Y, Zhao Y, Wang J, Zhang M, Jia W, Qin X. LDPE microplastic films alter microbial community composition and enzymatic activities in soil. Environ Pollut. 2019;254:112983.

Jang M, Shim WJ, Han GM, Cho Y, Hong SH. Plastic debris as a mobile source of additive chemicals in marine environments: in-situ evidence. Sci Total Environ. 2022;856:158893.

Jem KJ, Tan B. The development and challenges of poly (lactic acid) and poly (glycolic acid). Adv Ind Eng Polym Res. 2020;3(2):60–70.

Jemec Kokalj A, Dolar A, Drobne D, Marinšek M, Dolenec M, Škrlep L, et al. Environmental hazard of polypropylene microplastics from disposable medical masks: acute toxicity towards. Daphnia magna and current knowledge on other polypropylene microplastics. Microplast Nanoplast. 2022;2(1):1–15.

Jiménez-Skrzypek G, Hernández-Sánchez C, Ortega-Zamora C, González-Sálamo J, González-Curbelo MÁ, Hernández-Borges J. Microplastic-adsorbed organic contaminants: analytical methods and occurrence. TrAC Trends Anal Chem. 2021;136:116186.

Jin M, Yu X, Yao Z, Tao P, Li G, Yu X, Peng J. How biofilms affect the uptake and fate of hydrophobic organic compounds (HOCs) in microplastic: insights from an In situ study of Xiangshan Bay, China. Water Res. 2020;184:116118.

Kahn LG, Philippat C, Nakayama SF, Slama R, Trasande L. Endocrine-disrupting chemicals: implications for human health. Lancet Diabet Endocrinol. 2020;8(8):703–18.

Khan F, Ahmed W, Najmi A. Understanding consumers' behavior intentions towards dealing with the plastic waste: perspective of a developing country. Resour Conserv Recycle. 2019;142:49–58.

Koutnik VS. Microplastic accumulation and transport in the subsurface under weathering cycles. (Doctoral dissertation, UCLA); 2022.

Kumar P. Role of plastics on human health. Indian J Pediatr. 2018;85:384–9. https://doi.org/10.1007/s12098-017-2595-7.

Lahive E, Walton A, Horton AA, Spurgeon DJ, Svendsen C. Microplastic particles reduce reproduction in the terrestrial worm Enchytraeus crypticus in a soil exposure. Environ Pollut. 2019;255:113174. https://doi.org/10.1016/j.envpol.2019.113174.

Lehmann A, Fitschen K, Rillig M. Abiotic and biotic factors influencing the effect of microplastic on soil aggregation. Soil Syst. 2019;13:505–9.

Lei L, Liu M, Song Y, Lu S, Hu J, Cao C, et al. Polystyrene (nano)microplastics cause sizedependent neurotoxicity, oxidative damage and other adverse effects in Caenorhabditis elegans. Environ Sci Nano. 2018;5:2009–20.

Li J, Song Y, Cai Y. Focus topics on microplastics in soil: analytical methods, occurrence, transport, and ecological risks. Environ Pollut. 2019;257:113570. https://doi.org/10.1016/j.envpol.2019.113570.

Liu H, Yang X, Liu G, Liang C, Xue S, Chen H, et al. Response of soil dissolved organic matter to microplastic addition in Chinese loess soil. Chemosphere. 2017;185:907–17.

Liu H, Yang X, Liang C, Li Y, Qiao L, Ai Z, et al. Interactive effects of microplastics and glyphosate on the dynamics of soil dissolved organic matter in a Chinese loess soil. Catena. 2019;182:104177.

Lwanga E, Gertsen H, Gooren H, Peters P, Salanki T, van der Ploeg M, et al. Microplastics in the terrestrial ecosystem: implications for Lumbricus terrestris (Oligochaeta, Lumbricidae). Environ Sci Technol. 2016;50:2685–91.

Maaghloud H, Houssa R, Ouansafi S, Bellali F, El Bouqdaoui K, Charouki N, Fahde A. Ingestion of microplastics by pelagic fish from the Moroccan Central Atlantic coast. Environ Pollut. 2020;261:114194.

Made Safe. How to avoid toxic chemicals in plastics. 2016. https://madesafe.org/blogs/viewpoint/how-to-avoid-toxic-chemicals-in-plastics. Accessed Oct 2022.

Meng J, Zhang Q, Zheng Y, He G, Shi H. Plastic waste as the potential carriers of pathogens. Curr Opin Food Sci. 2021;41:224–30.

Mikhailovich K, Fitzgerald R. Community responses to the removal of bottled water on a university campus. Int J Sustain High Educ. 2014;15:330–42. [CrossRef]

Mngomezulu SK, Mbanga S, Adeniran AA, Soyez K. Factors influencing solid waste management practices in Joe Slovo Township, Nelson Mandela Bay. J Public Adm. 2020;55(3):400–11.

Mohanty AK, Wu F, Mincheva R, et al. Sustainable polymers. Nat Rev Methods Primers. 2022;2:46. (2022). https://doi.org/10.1038/s43586-022-00124-8.

Monkul MM, Özhan HO. Microplastic contamination in soils: a review from geotechnical engineering view. Polymers. 2021;13(23):4129.

Muscolo A, Settineri G, Attina E. Early warning indicators of changes in soil ecosystem functioning. Ecol Indic. 2015;48:542–9.

Napper IE, Bakir A, Rowland SJ, Thompson RC. Characterisation, quantity and sorptive properties of microplastics extracted from cosmetics. Mar Pollut Bull. 2015;99(1–2):178–85. https://doi.org/10.1016/j.marpolbul.2015.07.029. PMID: 26234612.

Ncube LK, Ude AU, Ogunmuyiwa EN, Zulkifli R, Beas IN. An overview of plastic waste generation and management in food packaging industries. Recycling. 2021;6(1):12.

Ndaw S, Remy A, Jargot D, Robert A. Occupational exposure of cashiers to bisphenol A via thermal paper: urinary biomonitoring study. Int Arch Occup Environ Health. 2016;89(6):935–46.

Nelms SE, Duncan EM, Broderick AC, Galloway TS, Godfrey MH, Hamann M, et al. Plastic and marine turtles: a review and call for research. ICES J Mar Sci. 2016;73(2):165–81.

Niyobuhungiro RV, Schenck CJ. The dynamics of indiscriminate/illegal dumping of waste in Fisantekraal, Cape Town, South Africa. J Environ Manag. 2021;293:112954. https://doi.org/10.1016/j.jenvman.2021.112954.

Nizzetto L, Bussi G, Futter MN, Butterfield D, Whitehead PG. A theoretical assessment of microplastic transport in river catchments and their retention by soils and river sediments. Environ Sci Process Impacts. 2016;18(8):1050–9. https://doi.org/10.1039/c6em00206d. PMID: 27255969.

Novikov MA, Gorbacheva EA, Prokhorova TA, Kharlamova MN. Composition and distribution of marine anthropogenic litter in the Barents Sea. Oceanology. 2021;61(1):48–57.

Onyedikachi UB, Belonwu DC, Wegwu MO. Human health risk assessment of heavy metals in soils and commonly consumed food crops from quarry sites located at Isiagwu, Ebonyi State. Ovidius Univ Ann Chem. 2018;29:8–24.

Onyedikachi UB, Belonwu CD, Wegwu MO, Ejiofor E, Awah MF. Sources and cancer risk exposure of polycyclic aromatic hydrocarbons in soils from industrial areas in Southeastern, Nigeria. J Chem Health Risks. 2019a;9(3):203–16.

Onyedikachi UB, Belonwu CD, Wegwu MO. Health risk assessment of chromium, manganese and arsenic through the consumption of food from industrial areas in south eastern states of Nigeria. Annu Res Rev Biol. 2019b;31:1–20.

Pedersen MW, Overballe-Petersen S, Ermini L, Sarkissian CD, Haile J, Hellstrom M, et al. Ancient and modern environmental DNA. Philos Trans R Soc B Biol Sci. 2015;370(1660):20130383.

Plastic Soup Foundation. The problem of plastic soup. Plastic Soup Foundation. https://www.plasticsoupfoundation.org/en/solutions/. Accessed 3 Feb 2023.

PlasticsEurope. Annual review 2017–2018. 2018. https://www.plasticseurope.org/download_file/force/1830/181.

Ragoobur D, Huerta-Lwanga E, Somaroo GD. Microplastics in agricultural soils, wastewater effluents and sewage sludge in Mauritius. Sci Total Environ. 2021;798:149326.

Rahman A, Sarkar A, Yadav OP, Achari G, Slobodnik J. Potential human health risks due to environmental exposure to nano-and microplastics and knowledge gaps: a scoping review. Sci Total Environ. 2021;757:143872.

Ramesh CH, Koushik S, Shunmugaraj T, Murthy MR. Mortality of sea turtles Chelonia mydas and Lepidochelys olivacea due to entanglement in fishing nets, in Mandapam region. Int J Curr Res. 2019;11(05):3660–2.

Ribeiro AS, Schoenfeld BJ, Souza MF, Tomeleri CM, Silva AM, Teixeira DC, et al. Resistance training prescription with different load-management methods improves phase angle in older women. Eur J Sport Sci. 2017;17(7):913–21.

Rillig MC, Lehmann A, de Souza Machado AA, Yang G. Microplastic effects on plants. New Phytol. 2019;223(3):1066–70. https://doi.org/10.1111/nph.15794.

Sayıcı IU, Simsek Orhon F, Topçu S, Ulukol B, Baskan S. Preliminary study on bisphenol A levels and possible exposure history of mother and exclusively breastfed infant pairs. Eur J Pediatr. 2019;178(4):541–50.

Selonen S, Dolar A, Jemec Kokalj A, Skalar T, Parramon Dolcet L, Hurley R, et al. Exploring the impacts of plastics in soil—the effects of polyester textile fibers on soil invertebrates. Sci Total Environ. 2020;700:134451.

Setälä O, Norkko J, Lehtiniemi M. Feeding type affects microplastic ingestion in a coastal invertebrate community. Mar Pollut Bull. 2016;102(1):95–101. https://doi.org/10.1016/j.marpolbul.2015.11.053.

Sharma R, Kaushik H. Micro-plastics: an invisible danger to human health. CGC Int J Contemp Technol Res. 2021;3(2):1–5.

Smith M, Love DC, Rochman CM, Neff RA. Microplastics in seafood and the implications for human health. Curr Environ Health Rep. 2018;5:375–86.

Song Y, Cao C, Qiu R, Hu J, Liu M, Lu S, et al. Uptake and adverse effects of polyethylene terephthalate microplastics fibers on terrestrial snails (Achatina fulica) after soil exposure. Environ Pollut. 2019;250:447–55.

Sridharan S, Kumar M, Singh L, Bolan NS, Saha M. Microplastics as an emerging source of particulate air pollution: a critical review. J Hazard Mater. 2021;418:126245.

Steinmetz Z, Wollmann C, Schaefer M, Buchmann C, David J, Tröger J, Schaumann GE. Plastic mulching in agriculture. Trading short-term agronomic benefits for long-term soil degradation? Sci Total Environ. 2016;550:690–705.

Sun Q, Ren S-Y, Ni H-G. Incidence of microplastics in personal care products: an appreciable part of plastic pollution. Sci Total Environ. 2020;742:140218. https://doi.org/10.1016/j.scitotenv.2020.140218.

Testai E, Buratti FM, Funari E, Manganelli M, Vichi S, Arnich N, Sialehaamoa A. Review and analysis of occurrence, exposure and toxicity of cyanobacteria toxins in food. EFSA Support Publ. 2016;13(2):998E.

Trasar-Cepeda C, Leiros M, Gil-Sotres F. Hydrolytic enzyme activities in agricultural and forest soils. Some implications for their use as indicators of soil quality. Soil Biol Biochem. 2008;40:2146–55.

United Nations. Department of Economic and Social Affairs Sustainable Development. The 17 goals; 2015.

Vandenberg LN, Hauser R, Marcus M, Olea N, Welshons WV. Human exposure to bisphenol A (BPA). Reprod Toxicol. 2007;24(2):139–77.

Verge X, De Kimpe C, Desjardins R. Agricultural production, greenhouse gas emissions and mitigation potential. Agric For Meteorol. 2007;142:255–69.

Wang J, Liu X, Li Y, Powell T, Wang X, Wang G, Zhang P. Microplastics as contaminants in the soil environment: a mini-review. Sci Total Environ. 2019;691:848–57. https://doi.org/10.1016/j.scitotenv.2019.07.209.

World Population Review. Plastic pollution by country 2022. 2022. https://worldpopulationreview.com/country-rankings/plastic-pollution-by-country. Accessed 19 Oct 2022.

Wright SL, Kelly FJ. Plastic and human health: a micro issue? Environ Sci Technol. 2017;51(12):6634–47. https://doi.org/10.1021/acs.est.7b00423.

Xiao X, Wang M, Zhu H, Guo Z, Han X, Zeng P. Response of soil microbial activities and microbial community structure to vanadium stress. Ecotoxicol Environ Saf. 2017;142:200–6.

Xu Z, Qian X, Wang C, Zhang C, Tang T, Zhao X, et al. Environmentally relevant concentrations of microplastic exhibits negligible impacts on thiacloprid dissipation and enzyme activity in soil. Environ Res. 2020;189:109892.

Ye X, Wong LY, Kramer J, Zhou X, Jia T, Calafat AM. Urinary concentrations of bisphenol A and three other bisphenols in convenience samples of US adults during 2000–2014. Environ Sci Technol. 2015;49(19):11834–9.

Yi M, Zhou S, Zhang L, Ding S. The effects of three different microplastics on enzyme activities and microbial communities in soil. Water Environ Res. 2021;93:24–32.

Zhou Y, Wang J, Zou M, Jia Z, Zhou S, Li Y. Microplastics in soils: a review of methods, occurrence, fate, transport, ecological and environmental risks. Sci Total Environ. 2020;748:141368.

Zhu B, Fang Y, Zhu D, Christie P, Ke X, Zhu Y. Exposure to nanoplastics disturbs the gut microbiome in the soil oligochaete Enchytraeus crypticus. Environ Pollut. 2018a;239:408–15.

Zhu D, Chen Q, An X, Yang X, Christie P, Ke X, et al. Exposure of soil collembolans to microplastics perturbs their gut microbiota and alters their isotopic composition. Soil Biol Biochem. 2018b;116:302–10.